Discover Long Island

Exploring the great places
from sea to sound

Barbara Shea

In memory of my parents, James and Emma Shea,
whose favorite places in the world were close to home.

Cover photo of the Fire Island Lighthouse by John Griffin. Other cover photos by Newsday.
Back cover photo of the St. James General Store by Newsday's Tony Jerome.

Printed by Interstate Litho Corp.,
Brentwood, N.Y.
ISBN: 1-885134-30-4

SPECIAL SALES: Newsday Books are available at discounts for bulk purchases or sales
promotions. For information on bulk and wholesale orders, please write to: Newsday Books,
235 Pinelawn Rd., Melville, N.Y. 11747

CONTENTS

Atlantis Marine World

Cradle of Aviation Museum

CONTENTS

Hempstead House, a Gold Coast mansion

Montauk Point Lighthouse

Planting Fields Arboretum

Deepwells Farm

CONTENTS

Sunken Meadow State Park

U.S. Merchant Marine Academy

Readers of this book are urged to call ahead before visiting any site. Fees, hours and dates of operation are subject to change. Resident passes are required for all Nassau County parks and preserves; call 516-572-0200 for information. Nassau County museums are open to the general public. Suffolk County parks are open to all, but resident and nonresident passes are available for discounts; call 631-854-4949.

Introduction

D O YOU KNOW where you can visit a French chateau, an Irish castle and an English manor all on one Gold Coast estate; stroll the path to enlightenment in a traditional Japanese garden; play tournament-level golf on a budget at the first municipal course ever to host the elite U.S. Open, and discover the secrets of countless international celebrities who've kicked back — and kicked up their heels — right in your backyard?

Neither did I, until Newsday asked me last year to take time off from roaming the globe for its Travel section, which has been part of my job for a dozen years, and turn a traveler's eye on the nearby world.

In 10 months, I drove some 3,500 miles on Long Island highways and byways from Kings Point to Montauk Point, zigzagging back and forth through time. I walked more than 200 miles along sandy beaches, wooded trails and village sidewalks; logged numerous nautical miles on local waterways, and toured nearly 300 museums, parks, wineries and historic sites, from colonial settlements to funky icons like the Big Duck — all within day-trip distance of any place in the New York metro area.

Though I've lived in both Nassau and Suffolk counties, I'd never visited most of these attractions. Perhaps like you, I'd even passed some every day on the way to work. At other spots I'd stopped at often, I was continually amazed at all that was new — or that I'd previously missed. How many beachgoers explore Fire Island's rare Sunken Forest, or climb its lighthouse? I hadn't.

A staggering number of Long Island's historical and natural attractions are literally national treasures, protected by landmark status. Some popular villages, such as Sag Harbor, have so many of these sites that the whole downtown has been declared a heritage area. Other destinations, whose day in the sun may not be widely remembered, are being reborn for new tourists. Creative recycling also has spawned everything from an award-winning nature preserve atop a reclaimed landfill to the ambitious museum row that's taking shape on a former airfield where aviation history was made.

At every turn, I encountered prominent Long Islanders both past and present (notably President Theodore Roosevelt, poet Walt Whitman and entertainer Billy Joel, who pop up everywhere). I learned, too, of astounding legacies left by less celebrated local heroes such as the slave acknowledged as America's first black poet and the woman whose simple landscaping idea has kept Montauk Point Lighthouse from toppling into the sea.

Most sites in this book are open year-round and are either free or inexpensive. Federal, state, county and some town parks departments issue annual passes or seasonal stickers that further reduce entrance fees. Many can be reached directly by train as well as car, or are a brief taxi ride from a Long Island Rail Road station. Almost all offer programs and events for children, adults and families — including those with physical limitations.

My favorite? Places, like friends, tend to fill different needs. I'd choose various natural areas for seasonal scenery, or for everyday solitude; certain parks and museums for fun activities, others for an educational sojourn. You'll soon develop your own list as you explore. Just take your time. Pause for a picnic or a restaurant meal, follow tangents and chat with people you meet along the way. You'll likely find that a single day's outing can be as refreshing as a distant journey — and often more so.

I've been truly bowled over by all there is to discover close at hand. I think you will be, too.

Newsday Photo / Ken Spencer

Barbara Shea

Visiting the John P. Humes Japanese Stroll Garden in Mill Neck

Discover
Long Island

ATLANTIS MARINE WORLD

Water Wonderland Gets a Family Seal of Approval

THOSE SOULFUL EYES. That cute little nose. Those jaunty flippers. With such an appealing group of harbor seals frolicking year-round in the front yard, who could resist a visit to Riverhead's new **Atlantis Marine World** to see what all the barking is about?

Take it from the whiskered welcoming committee — you won't be disappointed as you explore marine habitats from Long Island to the South Pacific and from Africa's Lake Malawi to South America's Amazon rain forest. You also can pet and help feed the touch-tank inhabitants; view X-rays — and usually a patient or two — at the **Riverhead Foundation for Marine Research and Preservation Marine Mammal and Sea Turtle Rescue Center**; shudder at the sight of a school of menacing-looking sharks; marvel at "the nation's largest public live coral reef" — as spectacularly colorful as the array of tropical fish parading around it in their latest designer garb; wonder how the 8-foot electric eels that can generate more than 600 volts to defend themselves aren't true eels but more closely related to goldfish; and search unsuccessfully for the winter flounder buried in the sandy bottom of their small tank — until you learn the trick of making the outline of the entire fish magically materialize.

A little theme park action? You can also take a simulated submarine dive

A harbor seal in the outside pool at Atlantis Marine World in Riverhead

Newsday Photos / Michael E. Ach

A sand tiger shark shares its tank with smaller species at the aquarium. Below, a lionfish, whose grooved spines can emit a potent venom

AT A GLANCE

Atlantis Marine World, 431 E. Main St., Riverhead; 631-208-9200 and www.atlantismarineworld.com. **Hours**: Open 9 a.m. to 6 p.m. daily Memorial Day to Labor Day, 9 a.m. to 5 p.m. the rest of the year (closed on Christmas). **Fee**: $12.50 adults, $10 ages 3 to 11 and over 62 (plus tax). Wheelchair accessibility; child appropriate.

Did You Know? Seahorse moms transfer their fertilized eggs into the dad's kangaroo-style brooding pouch, where he carries them until he "gives birth."

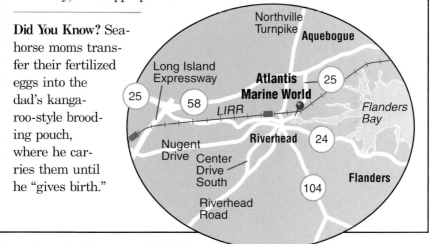

Take the Long Island Expressway to Exit 72S to Route 25. Follow Route 25 to West Main Street in Riverhead.

and, in the backyard, ride to the top of a 100-foot tower and clap along with some comical sea lions. And there's more — at last count Atlantis Marine World was touting some 80 exhibits.

The overall motif evokes, via giant murals and mock artifacts, mythology's legendary lost continent (any resemblance to the gigantic similarly themed aquarium at the Bahamas' Atlantis resort is pure coincidence, though, says curator and co-founder **Joseph Yaiullo**).

Riverhead visitors are welcomed to Poseidon's kingdom by that mythical god of the sea, whose ancient civilization supposedly disappeared beneath the waves after angering the sky gods. Halfway through an "underwater cavern" you find yourself staring-down more than a dozen steely eyed sharks — who share a 120,000-gallon tank with two creepy, 7-foot moray eels and assorted local and tropical fish (generally ignored by the well-fed "Jaws" gang). A viewing mezzanine offers an overhead vantage of their con-

tinuous aquatic ballet among the faux ruins of **Atlantis** — populated periodically by scuba divers.

In nearby tanks, a giant Pacific octopus broods, saucer-sized jellyfish
hover like aquatic UFOs and needle-toothed piranhas seem to eye the potential finger food outside the glass (actually not all these notorious canni-
bals are bloodthirsty monsters; some species are strict vegetarians, a sign-
board notes).

Newsday Photos / Michael E. Ach

A sea horse, of course, at the aquarium

Something friendlier? You can
safely pet the southern stingrays
and bamboo sharks in the **Ray
Bay** — reassured by a robotic
talking toucan perched overhead.

An aquarium highlight is the
sea lion show starring Jan, Free-
bie and Torey (who was featured
in the 1994 seal flick "Andre").
As long as it's mild enough for
these three hams to safely
emerge from the water at the outdoor amphitheater, they sing,
break-dance, do back flips and
"handstands," and kiss their
trainers on cue. (You get indoor
views of their pool year-round.)

Also in warm weather, visitors
can help **Cornell Cooperative Extension's** marine specialists gather
specimens on two-hour cruises of
the Peconic River and Flanders
Bay via the 75-passenger **Atlantis
Explorer.**

Want to be even more involved? You can sign up for an educational day
as a shark-keeper or seal trainer — or have your photo taken smooching a
sea lion or posing with aquarium mascots Jimbo Jaws or Oliver Octopus
(these last two are only humans costumed Disney-style). All these extras
naturally cost extra, and there's also a shop, of course, stocked with tons of
marine-themed toys and other gifts. Long-range plans include much more.
But already Atlantis Marine World is one truly fabulous fishpond. ◆

On Main Street, Star Confectionary — known as Papa Nick's to locals — was founded by Nicholas Meras in 1920. Still running it are his son Tony Meras, and grandson Anthony.

Hike, Drink and Be Merry

T EN MINUTES from **downtown Riverhead** you can pick peaches on a farm, sip chardonnay at a winery, hike through a piney wilderness, relax on a quiet beach, romp at one of the state's largest water parks, shop at a massive outlet mall and thrill to the action at one of America's oldest NASCAR racetracks.

But the walkable town center — long known best to Suffolk residents as the **"county seat"** where they go for jury duty, and to weekend sojourners as a quick snack stop en route to the East End — is becoming a why-go-anywhere-else-today destination itself. (The food and atmosphere at its many old-fashioned eateries, incidentally, *are* always worth a stop.)

In addition to the new Atlantis Marine World aquarium, there's the

Newsday Photo / Michael E. Ach

St. Isidore Roman Catholic Church, in the background, and Polish Hall at the site of Riverhead's annual Polish Town Street Fair & Polka Festival

Suffolk County Historical Museum (filled with 300 years worth of artifacts from the Indians and early pioneers to modern-day artists) and the **Railroad Museum of Long Island** (with a collection of vintage locomotives and cars). Venerable cultural forums include the 1881 **Vail-Leavitt Music Hall** (open for limited productions while undergoing restoration) and the thriving **East End Arts Council & Gallery** (which hosts a variety of concerts and other events).

As its name suggests, Riverhead also has a waterway — the lazy Peconic — which once irrigated its farms and powered its mills. Now the revitalized waterfront (don't miss the hand-carved figures topping the dock's pilings) attracts boaters, anglers, strollers and fun-seekers who flock to the annual **Blues Festival**, the **Country Fair** and the **Agricultural Heritage Festival**.

Many of Riverhead's early immigrants were Polish farmers who transplanted both their livelihood and culture — which still thrives in the neighborhood proudly proclaimed by signs and banners as Polish Town, USA. Every August, main-drag Pulaski Street (named for Gen. Casimir Pulaski, who died fighting with the Patriots in the American Revolution)

WHILE YOU'RE THERE

Robert Cushman Murphy County Park, off River Road in Manorville-Calverton, 631-854-4949. **Hours**: Open year-round. **Fee**: Free. Suffolk's largest park, and its first natural one, encompasses 3,084.9 acres around Swan Pond offering hiking, boating (free ramp), fishing and deer hunting in season (both with permit).

Cathedral Pines County Park, Yaphank-Middle Island Road, Middle Island, 631-852-5500. The 265-acre park has campsites (groups must reserve), playground, sports field, bridle paths, nature and mountain-biking trails and an activity center.

Prosser Pines County Park, Yaphank-Middle Island Road, Middle Island, 631-854-4949. Features a mile-long self-guided nature trail and good bird-watching on 56 acres.

Cupsogue Beach, west end of Dune Road, Westhampton Beach, 631-852-8111, off-season 631-854-4949. This county park offers limited outer-beach camping (no tents), seasonal amenities.

Southaven County Park, Victory Avenue, Brookhaven, 631-854-1414. **Fee**: Seasonal parking fee $2 residents (seniors and disabled free weekdays except holidays); $5 nonresidents. The 1,323-acre park has three miles of horseback riding trails (rentals), picnicking, camping, canoeing, rowing (rentals), freshwater fishing (fee, license), waterfowl hunting in season (license).

Terrell River County Park, Belleview Avenue access south of Montauk Highway, Center Moriches, 631-854-4949. **Hours**: Sunrise to sunset year-round. **Fee**: Free. This 260-acre site with year-round hiking trails is maintained by the Moriches Bay Audubon Society and Suffolk County.

For a sampling of Riverhead restaurants, see Page 292.

throngs with visitors who descend for the annual **Polish Town Street Fair & Polka Festival**. Everyone enjoys Old Country crafts and music, gobbles mouthwatering pierogi and babka — and for all practical (and frivolous) purposes is Polish for at least a day. ◆

Riverhead Town Hall, 631-727-3200 or www.riverheadli.com; Chamber of Commerce, 631-727-7600.

BELMONT PARK

Just
Horsing Around
At the Track

THEY'RRRRRE OFFFFF! The 9-year-old in the Yankees cap breaks fast, along with the pigtailed toddler riding her dad's shoulders, as the motley field of two dozen pretend-thorough-breds bursts from the starting gate amid giggles and grins.

It's a typical spring Sunday at historic **Belmont Park**, but hardly just another day at the races.

"Breaking" from an official starting gate — albeit not in front of the 70,000 who turn out on a Saturday afternoon each June for the Belmont Stakes — is all in a morning's fun at family-friendly **Breakfast at Belmont**, which features a behind-the-scenes look at the sport of kings before the track opens for the day.

Offered weekend and holiday race days May through July, September and October from about 7:30 to 10:30 a.m., Breakfast at Belmont affords visitors a chance to meet a jockey and stroke the velvety muzzle of one of the laid-back "ponies" that serve as a calming influence to the high-strung thoroughbreds. In addition, there's trackside commentary, equipment demonstrations and guided tram tours of the otherwise off-limits "backstretch" — not the track segment farthest from public view but the similarly named community of stable denizens, where the roads are named after horses and the stop signs say, "WHOA."

For a horse's-eye view, morning visitors to Belmont Park in Elmont can race out of the starting gate.

The Breakfast at Belmont program is free except for food, which can be purchased reasonably at the mustering point: the glass-walled **Belmont Cafe** in the clubhouse section of the viewing stands, which sprawl for nearly a quarter mile along the track's homestretch and finish line (but the elegant fourth floor **Garden Terrace Restaurant** overlooking the racecourse is the top choice for dining later in the day).

Whether sitting at an indoor table or standing out at the rail where you can feel the thunder of flashing hooves (and sometimes the sand they kick up), morning visitors get a panoramic view of the daily exercise routine. Some horses may be leisurely cantering, or "hacking," across the landscaped infield — where they're allowed only on Sunday mornings, to give them a chance to run on grass without tearing up the two turf tracks. Others may be "breezing" through timed workouts on the dirt main track — the outside ring of Belmont's three concentric racing ovals, and at 1.5 miles the longest in America. (At 430 acres, Belmont Park is also America's largest raceway.)

Newsday Photo / Paul J. Bereswill

Jockey Gary Stevens guides Point Given to a 12¹/₄-length victory in the 2001 Belmont Stakes. It was the race's largest margin of victory in 13 years.

After 11 a.m., when the track officially opens to horseplayers, there's a small admissions charge to either the grandstand or plusher clubhouse area for everyone age 13 and older. But family entertainment continues in **Belmont's "backyard,"** with everything from music to magic shows staged throughout the weekend afternoon races (children of any age are allowed at the track, but no one under 18 can bet).

In the garden-like paddock area behind the track building — which on this side resembles an old-world opera house, with four-story arched windows set into an ivy-covered brick wall — bettors scrutinize horses parading around the 175-year-old white pine that looks like a giant bonsai while kids eye the playground and picnic area. (TV monitors dot the yard so anyone not interested in puppet shows can view the races.)

Since the track's opening in 1905 (it was named for sportsman **August Belmont Jr.,** whose father gave *his* name to the stakes), as many horses as humans have left their mark there. The legendary Man o' War, defeated only once (by a horse named Upset), debuted at Belmont and easily took the 1920 stakes. **Secretariat**, which won the Belmont by a record 31 lengths to capture the 1973 Triple Crown, is honored by a paddock statue. As of 2001, only 11 horses had captured the Triple Crown with victo-

AT A GLANCE

Belmont Park, Elmont, 516-488-6000 or www.nyra.com. **Hours**: Belmont's spring-summer meet runs Wednesday to Sunday from the middle of May through late July; gates open for admission at 11 a.m. most race days (8:30 a.m. Belmont Stakes day); post time is usually at 1 p.m. (12:30 p.m. for Belmont Stakes; 3:05 p.m. Fridays for "Sunset Racing"). Belmont's fall meet runs from early September to late October; hours are the same, but there's no Friday sunset racing. **Parking**: $2 general, $4 preferred, $6 valet. **Fee**: General admission $2 (free Fridays for arrival before 1 p.m.), clubhouse $4 ($2 Fridays before 1 p.m.); ages 12 and younger free with an adult; reserved seats $2 to $10 depending on day and location. Partially wheelchair accessible. Child appropriate.

Did You Know? All thoroughbreds in North America officially become a year older on Jan. 1, even if they're born Dec. 31 (breeding is planned to avoid this).

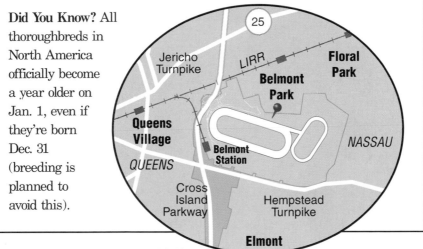

Take the Long Island Expressway to Exit 30
South onto the Cross Island Parkway. Go south about three miles to Exit 26D for Parking Field A, or follow signs for alternate parking options. If coming from the south, take the Southern State or Belt Parkway to the Cross Island Parkway north and follow signs for parking.

ries in the Kentucky Derby, Preakness Stakes and Belmont Stakes.

Photos of such historic highlights line the interior walls, and gift shops sell a range of racing memorabilia — and not all priced just for big winners. ◆

Horse Sense

HOW DO YOU give a pill to an ailing thoroughbred? Very carefully — via a foot-long, slim metal implement called a "balling gun," which horses reportedly also find more palatable (if less delectable) than an arm.

This is one of many intriguing bits of trivia you learn via Breakfast at Belmont. Among others:

● Thoroughbreds average 5 feet tall and weigh 900 to 1,200 pounds. Jockeys average 5 feet tall and weigh 100 to 110 pounds. In view of the disparity in brawn, jockeys must wear helmets and, under their colorful racing "silks," protective padded vests.

● Thoroughbreds can pack away up to 50 pounds of hay a day in addition to vitamins and other feed to maintain the stamina needed to speed along at 35 to 40 mph. No one seems to have documented the average jockey's

Newsday Photos / Bill Davis

A paddock show presenter at Belmont Park explains to visitors that a serial number tattooed inside its upper lip helps officials distinguish one thoroughbred from another.

diet, but it unquestionably makes them strong.

● Every thoroughbred is officially identified by its coloring and natural markings (such as white "stockings"), its callus-like leg growths called chestnuts (the equine equivalent of fingerprints) and a serial number tattooed inside the upper lip.

● Before a race, each horse is ID'd and examined to make sure it is physically fit. After a race, first- and second-place finishers (and sometimes others, if requested) are tested for drugs.

A horse fan introduces her 15-month-old son to a pony at Belmont's weekend paddock show.

● Trainers supply most of the horses' equipment, but jockeys buy their own miniscule saddle (which sometimes weighs as little as one pound). The jocks (saddle in hand) are "weighed out" before a race and "weighed in" at the end to ensure that all horses carry the proper ascribed poundage (in the Belmont Stakes, all horses carry the same weight).

● Since 1921, races on U.S. tracks have been run counter-clockwise (the reverse of England and some other countries). Even during workouts, horses aren't allowed to go "the wrong way" on the track unless they proceed slowly and stay to the outside fence.

● The starter pushes the button that springs the gate, but assistant starters go in with the horses to relax them and get them into position (which thoroughbreds should, but don't always, remember from the six-week starting-gate training school they must attend). How do starters know all the horses are ready? "When the gate gets quiet," an assistant explains, "and no one is yelling, 'No, no, no, no, no!' " ◆

WHILE YOU'RE THERE

If you don't plan to spend the whole day at Belmont Park, here are a few other sites to visit in the area:

Pagan-Fletcher Restoration, 143 Hendrickson Ave., Valley Stream, 516-872-4159. **Hours**: 1 to 4 p.m. Sundays (except holidays) and other times for special events. **Fee**: Free, except for some of the special events (call for information). This farmhouse, dating in part to the late 1700s, was home to Robert Pagan (credited with naming the village, said to be America's only Valley Stream) and later to his daughter and her husband, William Fletcher. It is on the National Register of Historic Places.

Valley Stream State Park, Southern State Parkway Exit 15 south, 516-825-4128. **Hours**: Open year-round, sunrise to sunset. **Fee**: $5 per car daily late May to early September; $5 weekends and holidays in early May and mid-September to mid-October. The smallest of the Island's state parks (97 acres) features picnicking, biking, hiking, boccie courts, basketball courts, playing fields and cross-country skiing in winter.

Hempstead Lake State Park, Southern State Parkway Exit 18 south, 516-766-1029. **Hours**: Open year-round, sunrise-sunset. **Fee**: $5 per car daily late May to early September; $5 weekends and holidays in early May and mid-September to mid-October. This 775-acre park, built around a sprawling lake, also offers a carousel. Facilities include playing fields, basketball courts, picnic areas, tennis courts, bike paths and bridle paths (nearby horse rentals).

Rock Hall Museum, 199 Broadway, Lawrence, 516-239-1157. **Hours**: 10 a.m. to 4 p.m. Wednesday to Saturday; noon to 4 p.m. Sunday. **Fee**: None. The museum was built in 1767 by Josiah Martin, a wealthy West Indian planter. It features an impressive example of Georgian architecture.

East Rockaway Grist Mill Museum, Woods and Atlantic avenues, East Rockaway, 516-887-6300. **Hours**: 1 to 5 p.m. weekends June to early September, or by appointment. **Fee**: None. This 300-year-old mill, burned by an arsonist in 1990, has been completely restored.

For a sampling of Elmont area restaurants, see Page 283.

Newsday Photo / Jim Peppler

Horseback riders have the trails to themselves on a snowy January day at Hempstead Lake State Park.

BETHPAGE STATE PARK

A Golf Course To Challenge The Best

H OW DID A state-owned golf course on Long Island get to host the 2002 U.S. Open — the first time in the prestigious tournament's 102-year history that it was held at municipal links?

Because it was time, said state Parks Commissioner Bernadette Castro. And because this wasn't just any public venue, but the top-rated Black Course at **Bethpage State Park** in Farmingdale — where foursomes often sleep in the car overnight for a shot at playing a $31 world-class round. Leaders of the U.S. and Metropolitan Golf associations had personal fond memories of "the Black," Castro said, and also "wanted to make a statement that the future of golf is public golf if it's to live on as a great American sport and one accessible to everyone regardless of economic background." Another likely catalyst: the huge success of a young phenom named Tiger Woods.

And so the **"People's Country Club"** got a $10-million face-lift to ready it for the "People's Open," which will leave it as a better, even more popular, destination for years to come. The cost was shared by the USGA, the state, the caterer and the pro shop. Primo golf architect **Rees Jones** waived his usual fee to oversee refurbishing of the 1936 course, which was designed by master **A.W. Tillinghast** and opened shortly after the park itself.

Tee time on the Black Course in Bethpage, the venue for the 2002 U.S. Open

The Black, one of an unprecedented five courses at Bethpage (the others are labeled Red, Blue, Green and Yellow) is for walkers only — no carts. The hilly, par-71 course (par 70 for the U.S. Open, when the seventh hole was designated a par 4 instead of 5) stretches 7,299 yards (the longest ever for an Open) and may be the only course anywhere to post a warning that recommends only highly skilled golfers dare challenge it. The legendary **Sam Snead** called it an "unfair test of golf" after defeating **Byron Nelson** there with a 68 in a 1940 exhibition match.

Which explains why with five 18-hole courses — what other end-of-the-round clubhouse bar is named the *91st* Hole? — golfers wait overnight for one of six first-hour walk-up slots or dial until their fingers fall off for a reservation (the res system does give priority to state residents). **The Black**, though closed Mondays and during the winter, is otherwise always there for them.

Golf may be the main attraction, but it isn't the only reason 1,500-acre Bethpage State Park logs some 800,000 annual visitors. Plenty of them agree with Mark Twain's quip that "golf is a good walk spoiled." (In the mid-1990s, local residents nixed a proposal to build the park's originally planned sixth course, which would have wiped out wooded picnic and wilderness areas.) The park is also popular for its seasonal weekly polo matches (Sunday afternoons mid-May to mid-October). The polo field is also used regularly for soccer matches and running events, with model airplane enthusiasts swooping in almost any time nothing else is going on there.

More traditional park activities include hiking (on nine miles of trails, including a piece of the **Nassau Suffolk Greenbelt Trail**), biking and roller skating (on a 1.4-mile park path — the start of the **Bethpage Bikeway**, extending south to Jones Beach), horseback riding (on more than four miles of bridle paths; rentals are available at private stables on Winding Road) and tennis (on four clay and four asphalt courts under two bubbles as well as four clay outdoor courts). Field games such as baseball and volleyball are favored by the 400,000 annual picnickers.

In winter, hills are set aside for supervised sledding and snowboarding, while certain trails are designated for cross-country skiing (but no snowmobiling).

This being such a surprisingly wooded park, given its midsuburbia location, it also harbors small wildlife. A sign behind the clubhouse points out that golfers may note birdies and eagles on their scorecard but can spot hawks, owls and other feathered friends, too. ◆

Newsday Photo / Nelson Ching

A championship polo match at Bethpage State Park

AT A GLANCE

Bethpage State Park, Bethpage Parkway, Farmingdale; general information 516-249-0700, golf reservations and information 516-249-0707, pro shop and golf lessons 516-249-4040. **Hours**: Park open dawn to dusk daily. **Fee**: Black course (no carts) $31 weekdays, $39 weekends; other courses $24 and $29 (carts available); all plus $3 for reservations. Park partially wheelchair accessible and child appropriate.

Did You Know? The park clubhouse was a Depression-era Work Relief Project; the "caddy boy" cutouts in its shutters were the idea of parkmeister Robert Moses.

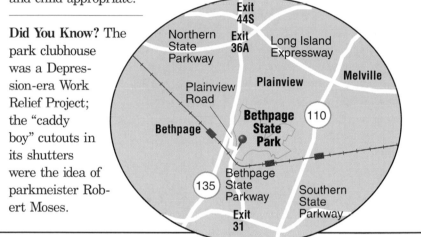

Take the Long Island Expressway to Exit 44S to Seaford-Oyster Bay Expressway (Route 135). Go south on Route 135 about four miles to Exit 8 (Powell Avenue). Go east on Powell (which becomes Plainview Road) a short distance to park entrance. From the south, take the Southern State Parkway to Exit 31 to Bethpage State Parkway north into the park.

◆

What a Dump!

CALL IT THE ULTIMATE in recycling — a town landfill reborn as an award-winning nature preserve.

Norman J. Levy Park & Preserve (still sometimes locally called Merrick Mountain) has also given Nassau County's South Shore a new landmark: a lofty windmill that stirs two wildflower-fringed ponds now gracing the 115-foot summit. Views from the 8-acre plateau encompass the

Manhattan skyline, New Jersey Palisades, Merrick Road Park Golf Course, Jones Beach water tower, Atlantic Ocean (and the Hempstead Town "resource recovery plant" that replaced the old landfill as the end of the line for nonrecyclables).

The 52-acre preserve, which opened at the end of 2000, also offers three miles of trails, 18 exercise stations, a 500-foot fishing pier (extending over prime **Merrick Bay** flounder, striped bass and bluefish territory) and a kayak and canoe launching ramp on revitalized **Meadow Brook** (yes, there actually is a stream flowing alongside the namesake parkway).

Foxes, rabbits, turtles, butterflies and dozens of bird species from owls to blue herons quickly took up residence among the poplars, cedars, phragmites, cattails and other vegetation. Some 50,000 cord grass seedlings were planted on six acres of reclaimed wetlands, but the other plants are "volunteers" seeded by the wind. All are growing vigorously (an advantage of establishing a park on top of a compost heap).

No toxic materials went into the site during its 34 years as a landfill (which ended in 1984), but the land, air and ground-

Newsday Photo / Michael E. Ach

A great blue heron flies by the Norman J. Levy Park & Preserve.

water were tested by the state **Department of Environmental Conservation** before it agreed to the park plan. The late Merrick state senator and environmental champion for whom the park is named would be pleased that benches and boardwalks are made of recycled materials, rest rooms have the latest composting toilets and the parking lot surface is crushed clamshells. Free jitney transportation around the 1.8-mile perimeter road is regularly available, guided riding and walking tours are offered, and there's an outdoor amphitheater for fair-weather nature classes. This proudly is a dump no more. ◆

Norman J. Levy Park & Preserve: On Merrick Road, just east of the Meadowbrook Parkway, turn south at sign for Hempstead Town Sanitation Department; 516-378-4210, ext. 379. Hours: Open 7 a.m. to dusk daily. Fee: Free.

WHILE YOU'RE THERE

Massapequa Historic Complex, Merrick Road at Cedar Shore Drive, 516-799-2023, www.massapequahistory.com, includes a book-filled 1896 library, a cozy circa-1870 servants cottage and the Gothic 1844 Old Grace Church (in its burial ground is the grave of Maj. Thomas Jones, the town's first non-Indian settler, for whom Jones Beach was named). Historical Society of the Massapequas volunteers can provide tales of the area's 19th century heyday as a popular resort — when prominent visitors included everyone from the duke of Windsor to sharpshooter Annie Oakley (for whom a street was named). **Hours**: Library year-round 10 a.m. to 1 p.m. Wednesdays and Saturdays. Church and cottage May through September, 2 to 4 p.m. Sundays (other times by appointment). **Fee**: Free.

Lauder Museum-Amityville Historical Society, occupies a former bank building at 170 Broadway, 631-598-1486. **Hours**: 2 to 4 p.m. Sunday, Tuesday and Friday. **Fee**: Donation. You'll find photos, research materials and memorabilia about the early days of this village that until the mid-1800s was known as South Huntington.

Cantiague Park, east of West John Street off Cantiague Rock Road, Hicksville, 516-571-7050 (tape) or 516-571-7056. The 115-acre Nassau County park has picnic and games areas, nine-hole golf course, mini golf, lighted driving range, indoor ice rink, water park.

Massapequa Preserve, between Merrick Road and Linden Street, 516-571-7443, is a 423-acre tract encompassing Nassau County's largest freshwater wetlands and largest pine barrens tract, 4.5 miles of Bethpage bikeway and five miles of Greenbelt hiking trail.

Newsday Photo / Bill Davis

Cantiague Park in Hicksville

Twin Lakes Preserve, Old Mill Road, Wantagh, 516-766-1580. The year-round Hempstead Town site offers 57 wooded acres with birdwatching and freshwater fishing (license required). **Fee**: Free.

For a sampling of Bethpage-Plainview restaurants, see Page 281.

CAUMSETT STATE HISTORIC PARK

Once It Was A Rich Man's Playground

T HE SIGN MARKING the turnoff promises a park, but all that seems to be at the end of the elegantly upswept entrance drive is a cluster of barns amid acres of open fields. No swings, no baseball diamonds, no campsites. This park looks an awful lot like a farm.

Actually it's both — and more.

Caumsett State Historic Park, on Huntington's Lloyd Neck peninsula, is the glorious remnant of a 1,750-acre summer estate developed in the early 1900s by publisher-philanthropist-department store heir **Marshall Field III**. Even back then, the spread — which encompasses woodlands, meadows and about two miles of rocky Long Island Sound shoreline — was more than a rich man's playground. It was a virtually self-sufficient community with upwards of 72 full-time gardeners alone, its own underground utilities (phone and electric lines as well as sprinklers), vast orchards, vegetable gardens and a dairy farm with 80 head of prize Guernseys (a four-acre walled garden near the parking lot is currently being restored).

Naturally, it also had the usual VIP accoutrements: yacht dock, polo fields, stables, kennels and pheasant coops to stock private

24

The old dairy barns at Caumsett State Historic Park on Lloyd Neck

hunts, as well as facilities for just about every sport except golf.

Today's 1,500-acre park, acquired by the state in 1961, may not offer visitors quite such a grand experience, but it does offer hiking, biking, jogging, bird-watching, cross-country skiing, fishing, scuba diving and occasional interpretive walks.

Because of its size — and its unpark-like lack of amenities, such as snack bars, as well as its ban on swimming, skating, skateboarding and pets — it also offers more solitude than many other state parks. The shore, for example, is a two-mile walk from the public parking area (closer access is allowed to anglers and divers by permit only, on a limited basis).

Caumsett's three-mile paved loop road is popular with binoculared, camera-toting walkers and bike riders precisely because it's off-limits to public vehicles. About 25 miles of sandy hiking trails also crisscross the grounds, which are filled with seasonal wildflowers and specimen trees (keep an eye out for the perfect gumdrop — actually a cluster of beech trees — on the left side of the road, about halfway between the parking lot and the main house).

Yes, there is a house — several, in fact, but only one that's open for tours (the **1711 Henry Lloyd Manor** on Lloyd Harbor Road, which served as the estate's main gatehouse in Field's time). Field's own Georgian-style mansion, where George Gershwin played piano and Fred Astaire danced for party guests, is now headquarters of the **Queens College Center for Environmental Teaching and Research** and not open to tourists.

But you can sit on a bench out back and enjoy the sweeping view of the Sound across the heart-shaped pond at the bottom of the hill. And at the east end of the backyard, you can peek into the Queens College Center's aviaries, housing owls, red-tailed hawks and bald eagles that have been permanently injured by accidents and can no longer survive in the wild. (You can also visit a similar **Volunteers for Wildlife** "hospital" near the park entrance, but please note the Field estate's polo pony barn is now part of an equestrian center open only to those boarding horses there or taking riding lessons.)

Newsday Photo / Michael E. Ach

A red-tailed hawk at the Volunteers for Wildlife rehab center at Caumsett

The name Caumsett, incidentally, was the Matinecock Indians' name for the area — meaning "place by a sharp rock." The boulder has long since been blasted away as a hazard to navigation (perhaps during World War II, when Fields allowed the government to monitor the Sound for enemy ships from his property). But rock hounds can combine a Caumsett Park visit with a stop at **Target Rock National Wildlife Refuge** down the road — which does still have its namesake boulder. ♦

AT A GLANCE

Caumsett State Historic Park, Lloyd Harbor Road (the continuation of West Neck Road), Lloyd Neck, 631-423-1770. Hours: Open sunrise to sunset daily. Fee: Parking fees in season. Partially wheelchair accessible. Child appropriate.

Did You Know? The Field house is a third smaller than it was originally — downsized, like many Gold Coast mansions, after income and property taxes were enacted.

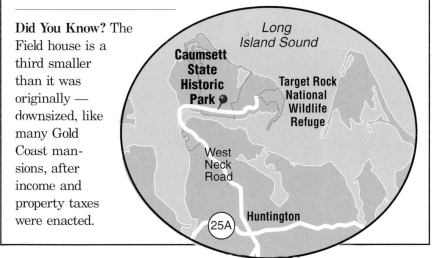

Take the Long Island Expressway to Exit 49N to Route 110 north. Take Route 110 approximately 10 miles to Main Street (Route 25A) in Huntington. Make a left on Main Street and go a short distance to West Neck Road on right. Follow West Neck Road about three miles until it becomes Lloyd Harbor Road. Caumsett Park is a short distance on the left; Target Rock is about another mile off Lloyd Harbor Road.

◆

Feudal Times on Long Island

A ROYAL LAND GRANT began the Lloyd family reign on the hilly land spit dense with lucrative firewood — the 18th century equivalent of 3,000 acres of oil wells. Before long, the peninsula was renamed **Lloyd Neck** from **Horse Neck** (the mystery is how early mapmakers deduced the equine profile without a helicopter flyover).

Slaves did much of the work on this plantation, but one named **Jupiter Hammon** was clearly favored through his tenure with three Lloyd generations. Not only was he among the few slaves who lived in the manor house (if in a small, unheated room), he was educated with the family's children, who called him "brother Jupiter." He was trusted to handle occasional business transactions — even allowed to accumulate money of his own. The Lloyds encouraged his interest in writing religious poems and helped him publish several prose and poetry works. His debut piece — **"An Evening Thought,"** dated Dec. 25, 1760 — has earned him the distinction as America's first published black poet.

Visitors learn more about these feudal times on Long Island via tours of the only remaining Lloyd houses: the red **1711 Henry Lloyd Manor** (protected during the Revolution because Henry II was a Loyalist) and the grander white **1766 Joseph Lloyd Manor House** (trashed by the British because this son was a Patriot).

Newsday Photos / Michael E. Ach

The Joseph Lloyd Manor House, built in 1766, retains many of its original features.

Both houses were modified over the decades but retain many original features. More recently, aviator **Charles Lindbergh** lived in the Joseph Lloyd House for about two years to escape the limelight after his baby was kidnapped. And **Marshall Field** used the Henry Lloyd Manor as the gatehouse to his Caumsett home.

The walled, waterfront estate on the curve at the end of the Lloyd Neck causeway was also a gatehouse, by the way — for a hilltop mansion on the former site of a British fort named for Benjamin Franklin's estranged illegitimate, Loyalist son, William (and you think politics today is scandalous). But the estate is perhaps most famous as a onetime home of singer **Billy Joel**. It's said he moved because boaters kept cruising by to ogle his then-wife: model Christie Brinkley. ◆

Henry Lloyd Manor, 41 Lloyd Harbor Rd., 631-424-6110. Hours: Open Saturday and Sunday noon to 5 p.m. year-round. Joseph Lloyd Manor House, Lloyd Harbor Road and Lloyd Lane, 631-692-4664. Hours: Open Saturday and Sunday 1 to 5 p.m. late May through mid-October.

WHILE YOU'RE THERE

Target Rock National Wildlife Refuge on Lloyd Harbor Road, Lloyd Harbor, 631-286-0485. **Hours**: Open daily from a half hour before sunrise to a half hour after sunset. **Fee**: $4 per vehicle, $2 bicyclist / pedestrian, $12 annual pass. The refuge occupies a former estate now converted to an 80-acre oak-hickory forest. In spring, the woods are splashed with the yellow and pink blooms of daffodils, azaleas and rhododendrons. A 10-minute walk west along the beach from where the three-quarter-mile trail emerges from the forest takes you to the 14-foot-high boulder, now in the water after centuries of shore erosion (and like Massachusetts' Plymouth Rock, looking a bit smaller than its legend).

Target Rock at Lloyd Neck

British seamen reportedly practiced marksmanship on the rock during the Revolutionary War and the War of 1812; the U.S. Navy had its shot at it during a Fourth of July demonstration in 1879.

Van Wyck-Lefferts Tide Mill in Lloyd Harbor is the only one of a handful of similar East Coast structures that's on its original site, complete with its wooden gear works. It was built by Abraham Van Wyck between 1793 and 1797, then sold to Samuel and Henry Lefferts in 1798 — which began its seven decades of operation. The mill, which got its power from tidewater receding from the mill pond, stands on a Nature Conservancy sanctuary now surrounded by private property, and is only accessible to the public by boat. About a dozen excursions are offered each year — half sponsored by the conservancy, half by the Huntington Historical Society. For more information, call 631-367-3225.

For a sampling of nearby restaurants, see Cold Spring Harbor on Page 282 and Huntington on Page 287.

COLD SPRING HARBOR

A Main Street Swimming With History

TINY **COLD SPRING HARBOR**, almost 100 miles from the open ocean, was a doomed player in the 19th century whaling game. Its nine ships valiantly pursued the lucrative beasts from the South Pacific to the Arctic, but were no match for the vast fleets from major ports like Nantucket. But the Long Island village gives the Massachusetts resort island a run for its money today when it comes to charming Main Street attractions.

During whaling's circa-1850 heyday, Cold Spring Harbor's main drag was nicknamed **Bedlam Street** because of the cacophony of foreign tongues (and colorful language) heard there when the ships were in. But the international crews, which kept changing as voyages crisscrossed the paunch of the globe, generally worked together harmoniously — giving whaling the reputation as the first integrated business. Investors in Cold Spring (as it was called then) barely broke even financially, but they may have set the stage for the worldly North Shore community it was to become. In the North Shore's Gold Coast years, **Louis Comfort Tiffany** had a hilltop estate overlooking the harbor and immortalized local views in stained glass.

The **Cold Spring Harbor Whaling Museum** exhibits thousands of objects and documents pertaining to the whaling era (which locally lasted from 1836 to 1862) and whale conservation (while the United States was once the

Main Street in Cold Spring Harbor; below, a scene at the whaling museum

world's most important whaling country, it no longer participates in any commercial whaling).

In the main gallery you can scrutinize a diorama of the harbor in 1850 and run your hand along New York's only fully equipped 19th century whaleboat — a six-man 30-footer pulled on many a wild **"Nantucket sleigh-**

ride" by a harpooned whale trying to escape its fate. Nearby is a model of its mother ship, the brig Daisy, built in Setauket in 1872 (and lost at sea in 1916). Next door in **The Wonder of Whales** room, kids especially love to listen to the "songs" of humpback whales and quiver before the skull of a killer whale (the only species with the nerve to attack other whales) and a sperm whale jaw (its formidable tines technically not teeth, though victims probably don't care whether they're about to be chewed up or bolted down whole).

To while away their spare time on voyages that lasted up to five years, whalers engraved scenes on the only part of the whale that wasn't other-

Newsday Photo / Barbara Shea

The Cold Spring Harbor Whaling Museum

wise used: the teeth and bones. This art, known as scrimshaw, was applied to utilitarian items such as sail makers' tools and kitchen gadgets, as well as to play items such as dominoes, personal trinkets such as necklaces, and larger decorative objects from vases to lamps. The museum has a renowned collection of some 700 such pieces in its permanent collection, which also includes marine art, ship models and figureheads. Also featured are special programs and village walking tours.

Numerous homes and commercial buildings dating to the whaling days line the quarter-mile **"Captain's Row"** business district, where you can buy anything from chocolate bonbons to an antique blanket chest.

Other Main Street attractions include the **Society for the Preservation of Long Island Antiquities Gallery** (offering changing exhibits and programs), **Cold Spring Harbor Laboratory's DNA Learning Center** (offering a film on Long Island history and a soon-to-reopen biotechnology museum), a state park, a marina, plus nearby nature preserves and a historic fish hatchery. Oh, and a waterfront village park named for Long Island's homegrown singer-songwriter Billy Joel — whose first album, fans will remember, was titled "Cold Spring Harbor." Top that, Nantucket. ◆

AT A GLANCE

Cold Spring Harbor Whaling Museum, Main Street, 631-367-3418 or www.cshwhalingmuseum.org. **Hours**: 11 a.m. to 5 p.m. Tuesday to Sunday (daily in summer). **Fee**: $3 adults, $2 seniors, $1.50 ages 5 to 18. Museum wheelchair accessible and child appropriate; village partially wheelchair accessible.

Did You Know? Blue whales, more than 100 feet long, are the largest animals ever to inhabit Earth; smaller, slower "right whales" were considered the right ones to hunt.

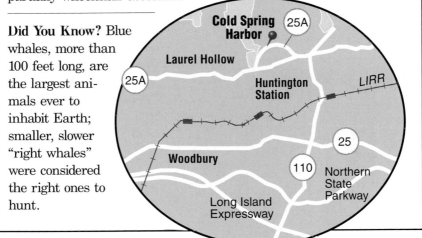

Take the Long Island Expressway to Exit 49N to Route 110 north. Take Route 110 about eight miles to Route 25A in Huntington. Make a left (west) onto Route 25A. Continue west on 25A for about three miles to Main Street.

◆

Hooked on Fun

WHAT OTHER fish hatchery holds an annual birthday party complete with cake (for the humans) and the requisite celebratory song (for the trout) to welcome each year's new batch of 40,000 hatchlings?

And that's only January's entry on the annual calendar of events at the **Cold Spring Harbor Fish Hatchery & Aquarium** (which is technically in Laurel Hollow). There are also spring and fall fishing days when kids have 10 minutes to try their luck (catches can be released or taken home), a Turtle & Tortoise Pageant on Herp Day (along with exhibits by the **New**

York Herpetological Society) and other informational and fun stuff.

The hatchery-aquarium says that the slimy and scaly creatures inhabiting its indoor and outdoor tanks and pools comprise the largest living collection of New York State freshwater fish, reptiles and amphibians.

Founded as a state hatchery in 1883, when the first brown trout imported into the United States arrived there, it switched gears in 1982 and became a private nonprofit **"aquatic environmental education center and demonstration hatchery"** raising brook, brown and rainbow trout that are sold to stock private ponds. Visitors usually also can purchase a trout or two for supper, and can always buy food to feed the fish.

Newsday Photos / Michael E. Ach

Brook trout at the fish hatchery

Because trout reared at the hatchery can't spawn naturally without gravel stream beds in which to deposit their eggs, the process proceeds artificially. In early November, females are "stripped" of their eggs and males of their "milt."

Visitors can watch as hatchery staffers and volunteers standing hip-deep in the fish ponds perform the quick task at scheduled times. The next step occurs in "a plain old silver bowl." Fertilized eggs then go into a **"hatch house"** where they're protected from predators and sunlight for about six weeks. Hatchlings stay there four months before moving on to rearing pools, then larger ponds.

A warm-water pond holds an assortment of bass, catfish, carp, pike and other species kept as an educational display. There's also a turtle pond, whose occupants are released into proper habitats when they're a year old. But if you want a turtle to call your own, visitors are told, you'll have to go to a pet store. ◆

Cold Spring Harbor Fish Hatchery & Aquarium, Routes 25A and 108, 516-692-6768 or on the Internet at www.cshfha.org. Hours: 10 a.m. to 5 p.m. daily. Fee: $3.50 adults, $1.75 ages 5 to 17 and over 65.

WHILE YOU'RE THERE

Dolan DNA Learning Center, 334 Main St., 516-367-5170 or on the Internet at http://vector.cshl.org. **Hours**: The educational arm of Cold Spring Harbor Laboratory is revamping its biotechnology museum, but shows Cablevision's "Long Island Discovery" multi-image surround sound take on local heritage to the general public at 1 and 3 p.m. Saturdays during the school year and at 1 and 3 p.m. Monday through Saturday in summer. **Fee**: Free.

Society for the Preservation of Long Island Antiquities Gallery,

Billy Joel Cold Spring Harbor Park

Main Street and Shore Road, 631-692-4664. Varying hours and exhibits. **Fee**: Free.

Billy Joel Cold Spring Harbor Park is a small waterside park off Route 25A that in 1991 was renamed to honor Long Island's own star singer, songwriter and champion of the environment.

Cold Spring Harbor State Park, Route 25A, across from the harbor, call Caumsett State Park, 631-423-1770. **Hours**: Open year-round sunrise to sunset. This recent addition to the state parks system offers 40 acres of wooded hillside with scenic views and the northern trailhead of the 22-mile Nassau Suffolk Greenbelt Trail. Cross-country skiing and snowshoeing areas are available in winter.

Uplands Farm Sanctuary, 250 Lawrence Hill Rd., 631-367-3225 or on the Internet at www.nature.org/longisland. Two trails loop through fields and woods on the 97-acre property of The Nature Conservancy's Long Island Chapter, which has its headquarters in the farm buildings. (The conservancy's St. John's Pond Preserve adjacent to the fish hatchery offers trails on 14 acres of marsh and woodland; pick up gate key at the hatchery.)

For a sampling of nearby restaurants, see Cold Spring Harbor on Page 282 and Huntington on Page 287.

CONNETQUOT RIVER STATE PARK
AND BAYARD CUTTING ARBORETUM

Purely Natural On The South Shore

S HERWOOD FOREST meets the Vienna Woods at two adjacent nature preserves in the heart of the South Shore's 19th century Gold Coast.

At **Connetquot River State Park Preserve** in Oakdale, hikers half-expect England's legendary Robin Hood to gallop from behind every stout oak. But only an occasional local horseback rider breaks the mood of a forest primeval in the 3,473-acre preserve — a onetime private hunting ground that now scrupulously protects its bountiful flora (including wild orchids) and fauna (deer, wild turkeys and more than 200 species of birds) and restocks its waters for controlled fishing revered by anglers from around the world.

Austria's more manicured woods are evoked at adjoining **Bayard Cutting Arboretum** in Great River, where brawny weeping beeches and feathery pines shade wildflower gardens and sweeping lawns that surround an Americanized Tudor-Queen Anne (and remotely alpine) mansion. The 697-acre arboretum, which also borders the river, is especially noted for spring-flowering shrubs such as rhododendrons and azaleas.

Both properties acknowledge they're not your usual state parks. From the

Daffodils in bloom behind the main house at Bayard Cutting Arboretum

long list of prohibitions (Bayard specifies no pets, bikes, picnics, Frisbees, kites, jogging, ball playing, climbing trees, picking flowers; Connetquot adds no feeding of wild animals and no smoking — plus requires a permit), you may conclude they're no fun. But they're meant to introduce suburbanites to nature at its purest — and to show how blissful this can be.

Railroad and shipping tycoon **William Bayard Cutting** began developing the arboretum in 1887 at his summer estate — one of many built around that time on the South Shore by wealthy men less known than the titans of industry who a generation later firmly affixed the Gold Coast label to the North Shore. Part of Bayard's estate was designed by the renowned firm of **Frederick Law Olmsted**. Though many of its oldest, tallest evergreens were destroyed by Hurricane Gloria in 1985, the remaining collection is still thought to be the most extensive on Long Island. And up close, you'll see that the giant weeping beech in the side yard has served as more than a natural umbrella. A wooden deck now makes it easier to walk around the

Newsday Photos / Tony Jerome

An autumn scene at Bayard Cutting Arboretum

rooty trunk but harder to carve any more graffiti into its papery bark (beech comes from an old English word meaning book of the woods, and this one is a veritable encyclopedia).

Only a few rooms in the house are open to view, but are worth a visit for the antique fireplaces, **Tiffany** windows and great gift shop. There are lawn concerts most summer Sundays and frequent guided walks.

Across the meandering river (and headlong Sunrise Highway), you enter a wilder world at the Connetquot preserve, where the roar of traffic soon morphs into the burbling of water and twittering of birds. From 1866 to 1973, this was the province of the **South Side Sportsmen's Club**

Connetquot's fish hatchery raises trout in holding ponds.

AT A GLANCE

Bayard Cutting Arboretum, Montauk Highway, Great River; 631-581-1002. **Hours:** 10 a.m. to sunset Tuesday to Sunday plus legal holidays year-round. **Fee:** There's a $5 parking fee daily April to Labor Day; free at other times. Wheelchair accessibility. Child appropriate.

Connetquot River State Park Preserve, Sunrise Highway, Oakdale; 631-581-1005. **Access is only by free permit**, valid for one year (write to Box 505, Oakdale, N.Y. 11769); added requirements for fishing. For information on guided walks and other programs, call 631-581-1072. **Hours:** Sunrise to sunset Tuesday to Sunday, April through September; Wednesday to Sunday October through March. **Fee:** Admission is free but there's a $5 parking fee year-round. Partial wheelchair accessibility. Child appropriate.

Did You Know? The Connetquot is designated a New York State Recreational River for the stretch through Connetquot River State Park Preserve, attesting to the high water quality (thus fishing is allowed, but not canoeing).

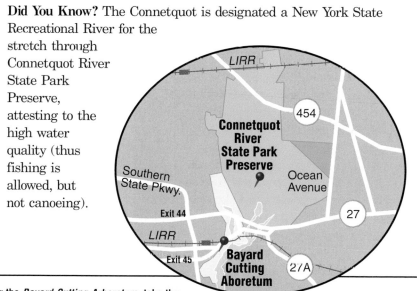

For the *Bayard Cutting Arboretum*, take the Long Island Expressway to Exit 53 south to the Sagtikos State Parkway. Take the Sagtikos south to the Southern State Parkway east. Southern State will become Heckscher Spur Parkway. Take this south to Exit 45 east to Montauk Highway (Route 27A). The aboretum is on the right. For the *Connetquot River State Park and Preserve*, take the Southern State Parkway east to Exit 44, then follow signs to the preserve.

Newsday Photo / Tony Jerome

A swan is undeterred as fishermen work their lines at Connetquot River State Park Preserve.

— whose illustrious members included President Grover Cleveland and Gen. William Tecumseh Sherman. The circa-1820 clubhouse building by the pond (where there's also an 18th century gristmill) was initially a stagecoach stop called **Snedecor's Tavern**.

About a mile's walk through woods and fields (only seniors and those with medical reasons may drive) is the hatchery, where fingerlings are raised in outdoor holding ponds covered with netting. But don't be surprised if a trout drops out of the sky; sometimes a sly osprey snatches, then loses, its wiggly prey. To satisfy young visitors' desire to feed something, a dispenser is stocked with pellets for the ducks plying the canal.

Want to fish? The species are limited to **brook, brown** and **rainbow trout**. But thousands are released each year and about 12,500 anglers book appointments to fly-cast from boats or sites with platforms (which avoid damage to stream banks).

The preserve also offers a variety of intriguing year-round programs for children, adults and families. Even Robin Hood likely couldn't resist a "bat safari." ◆

Wildlife in the Winter

WHEN WINTER unfurls a white blanket over the parks, even familiar trails take on a different look. Camera buffs, especially, wouldn't miss an early morning visit after a snowfall — when pines bow more gracefully than ever (and a mischievous squirrel may shake dabs of hemlock frosting down your neck).

At **Bayard**, photographers are drawn to the graceful weeping beeches, whose silvery tentacles gain extra sparkle from the lightest dusting. Surprising contrast to bare branches and evergreens is provided by an elderly beech hedge curiously flecked with brown all winter (the two annual trimmings needed to tame its growth also throw off its metabolism, but the old leaves never fail to yield to new spring sprouts).

Connetquot, though not Bayard, also allows cross-country skiing but along the river in both parks anyone can spot hooded mergansers, canvasback and wood ducks,

Newsday Photo / Thomas R. Koeniges

In a scene reminiscent of a holiday card, deer and geese gather in the snow at Connetquot River State Park Preserve.

great blue herons, pied-bill grebes and other migratory waterfowl that spend warmer months farther north and stop here to feed at unfrozen pockets of water. As always, there are Canada geese and swans.

Nocturnal animals such as foxes and raccoons remain elusive, and white-tailed deer skittish, but winter days offer plenty of curious chipmunks. It's also fun to follow in the tracks of those animals you never see. Don't worry — there are no bears. ◆

Newsday Photo / John H. Cornell Jr.

Skaters enjoy a sunny March day at Heckscher State Park.

WHILE YOU'RE THERE

Heckscher State Park, at southern end of Heckscher Spur Parkway in East Islip, 631-581-2100. **Hours**: Vary by season; open year-round. **Fee**: $7 per car daily late June through Labor Day; $5 weekends April to Memorial Day, and then daily to late June; $5 weekends and holidays after Labor Day through early October. This nearby park offers beach and pool swimming, hiking, biking, ball playing, camping, cross-country skiing, boat ramp, concerts and activities not permitted in adjoining state preserves.

Islip Art Museum, Brookwood Hall Park, 50 Irish Lane, East Islip, 631-224-5402. **Web**: www.islipartmuseum.org. **Hours**: 10 a.m. to 4 p.m. Wednesday to Saturday, noon to 4 p.m. Sunday. **Fee**: $2 donation. The museum specializes in avant-garde art, with about seven exhibits yearly, a permanent collection, talks and seminars plus the Carriage House, an experimental work space. Focus is on Long Island and Manhattan art.

Anthony Giordano Gallery, Idle Hour Boulevard, Oakdale, 631-224-3016. **Hours**: 10 a.m. to 4 p.m. Wednesday to Saturday, noon to 4 p.m. Sunday. **Fee**: None. This gallery at Dowling College has contemporary exhibits featuring local and nationally known artists. Schedule includes student and faculty exhibits.

For a sampling of nearby restaurants, see Bayport-Patchogue area on Page 280 and Sayville on Page 293.

Newsday Photo / David L. Pokress

Winds on the Great South Bay propel sailboards at Heckscher State Park.

CRADLE OF AVIATION MUSEUM

A Flight Through Time And Space

O VERHEAD, a vintage Navy fighter jet hangs, mid-dive, in a surreal sky that also includes a historic biplane, a space-walking astronaut and a helicopter that looks like a cross between a go-cart and a propeller beanie.

And that's just immediately inside the door of the new **Cradle of Aviation Museum** — centerpiece of an ambitious museum row taking shape in Garden City. The **Long Island Children's Museum** has already moved into its innovative new quarters at the Museums at Mitchel Center complex, and long-range site plans also include a **Nassau County Firefighters Museum & Safety Center**, a science and technology museum, and an events plaza with the Island's landmark **Nunley's Carousel**.

But of all the attractions slated to breathe new life into renovated hangars at this former airfield, none is more fitting than the Cradle of Aviation Museum (scheduled to open in May, complete with Long Island's first domed IMAX theater). The dozens of rare craft already in its continually growing collection represent every key period in aerospace history — which in the first two-thirds of the 20th century zoomed from rickety cracker boxes to high-tech spaceships. What's more, all those on display were either manufactured on Long Island or have some significant local connection. The museum's **Lunar Module**, built by Grumman Corp. for the Apollo pro-

A Fleet biplane and a Grumman F-11 Tiger in Blue Angels colors hang in the atrium of the Cradle of Aviation Museum at Mitchel Field.

Newsday Photos / Ken Spencer

A mannequin of journalist Harriet Quimby, who trained on Long Island to become the first licensed female pilot, with a Bleriot XI monoplane

gram, is one of only three LEMs remaining on Earth.

While the Wright brothers' pioneering flights lifted off from North Carolina, a staggering number of aviation firsts are linked to the New York metro area — primarily Nassau County's flat-as-a-Texas-prairie **Hempstead Plains**. Most famous was Charles Lindbergh's 1927 nonstop transatlantic solo in the single-engine Spirit of St. Louis — which started from **Roosevelt Field** (before it became a shopping mall). From the late 19th century, when daredevils began jumping off North Shore cliffs in birdman-like gliders, to the early days of aviation's 1918-to-1939 Golden Age, when about 70 airfields dotted Long Island, the countless other records set here include the first radio transmission from a plane (and subsequent first aerial traffic reports), first official airmail flight (a short hop from Garden City to Mineola) and first nonstop cross-country air journey (a previous attempt taking 49 days involved so many crashes that only two original parts were left — which explains why the museum has only a replica of *that* plane).

The Hempstead Plains also had the first U.S. civilian flying school,

AT A GLANCE

Cradle of Aviation Museum, 1 Davis Ave. (off Charles Lindbergh Boulevard), Garden City, 516-572-0411 or www.cradleofaviation.org. Museum hours and fees hadn't been finalized when this book went to press, so call before you visit. **Hours**: Museum hours were expected to be 9 a.m. to 5 p.m.

The glass-front entrance to the Cradle of Aviation Museum adjoins hangars

daily. **Fees**: $6 adults and $4.50 children for museum entrance, $8.50 and $6 for the IMAX theater, with various combination rates anticipated. Wheelchair accessibility. Child appropriate.

Did You Know? Roosevelt Field was named for Theodore Roosevelt's son Quentin, who trained at the air base there for World War I service and was killed in action over France in 1918. Adjacent Mitchel Field was named for New York City Mayor John Purroy Mitchel, who died the same year in a training accident in Louisiana.

From the Northern State Parkway take Exit 31A south to the Meadowbrook Parkway. Get off Meadowbrook Parkway at Exit M4 and go to Charles Lindbergh Boulevard. Turn right at the second traffic light into the museum parking lot. From the Southern State Parkway take Exit 22 and go north on the Meadowbrook Parkway to Exit M4 and follow directions above.

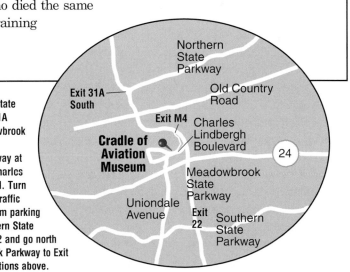

Newsday Photos / Ken Spencer

The Lilienthal glider, one of the first successful flying devices

whose graduates included journalist **Harriet Quimby** (first licensed American woman pilot) and flying ace **Jacqueline Cochran** (depicted, white scarf blowing, in the biplane she learned to fly in — which now hangs in the museum's four-story glass atrium visitor center).

Among other historic treasures that have been restored by some 140 volunteers (knowledgeable docents will be on hand to answer visitors' questions) are a **1909 Bleriot** that's America's fourth-oldest aircraft, the **1918 Curtiss JN-4 "Jenny"** that was Lindbergh's first plane (he bought it after only a few flying lessons and soloed it home) and one of two surviving 1928 sister ships of the Spirit of St. Louis (Lindy flew the reproduction around Roosevelt Field only for fun during filming of the 1957 tribute to him; the movie, named for the plane, starred actor James Stewart).

The museum will hit chronological highlights — via displays including actual aircraft, replicas, films and eventually also flight simulators and additional site mock-ups — from the times of hot-air balloons and early "air meets" through two world wars, the jet age and periods of space exploration to the present day. (Did you know Long Island has had six astronauts?)

Also on site are a gift shop and the **Red Planet Cafe**, whose decor simulates a 21st century Mars space base. And that's just Phase I. Among other museum-related features planned: aviation and space camps. Aviation's golden age may now be history, but local enthusiasts will always continue reaching for the stars. ◆

A structure of undulating platforms, ClimbIt! invites all to reach new heights at the Long Island Children's Museum.

Imagination on Display

SUPPOSE YOU COULD jump inside a giant toy box filled with playthings that would give you magical powers. You'd be able to build anything you can imagine, from sand dunes to skyscrapers. Create a kid-sized soap bubble around yourself — and learn the trick to poking it with a finger (or nose) without popping it. Be an actor, a musician or TV broadcaster. Climb an awesome aerial structure higher than your wildest dreams.

This only begins to describe the new **Long Island Children's Museum**, the first attraction to open on Garden City's budding Mitchel Center museum row.

It's a museum with a message — multiple messages, actually. A primary one is that the journey through childhood should be fun as well as educational. Another is that anything under the sun can be recycled — a perfect example being the museum's own quarters in a renovated airplane hangar now sporting bright colors and a dozen interactive computer-dotted exhibit areas.

The two-story play structure called **ClimbIt!** seems to float beneath the roof like a multi-tiered cloud bank held together by spider webs (really sturdy protective cables). An undulating ramp weaving through it is designed to deliver another key message: The sky's the limit for every kid — even those in wheelchairs.

TotSpot, the preschool area, has a soft "baby boat" with seagull mobiles and shore sounds to engage newborns — plus a toddler-sized town with shops, vehicles, a lighthouse, a sponge-rock jetty and even a construction site. Adjacent are a learning studio for messy projects such as finger painting, a quiet nook where adults can read to little ones and a resource room with staff ready to help solve sticky parenting problems from finicky eating to sibling rivalry.

There's a 150-seat theater where kids can produce as well as attend shows, a gift shop with educational toys and a gallery intended for continually changing displays (the initial one shares drawings and dreams that went into creating the children's museum).

Most exhibits target kids up to age 12, but this is a place so full of grand surprises it's going to be hard to keep grown-ups from hogging some of the fun stuff. ◆

Long Island Children's Museum, 11 Davis Ave. (off Charles Lindbergh Boulevard), Garden City, 516-222-0207 and www.licm.org. Hours: Open 10 a.m. to 5 p.m. Wednesday to Sunday. Fee: $8 for everyone over age 1.

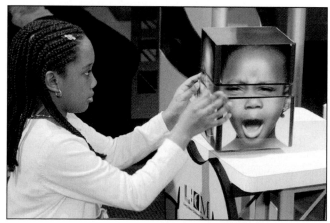

An exhibit with three rotating segments shows how we communicate with facial expressions.

Newsday Photo / Ken Spencer

Newsday Photo / Alejandra Villla

A 1925 Harley-Davidson at the Nassau County Police Museum

WHILE YOU'RE THERE

Nassau County Police Museum, 1490 Franklin Ave. (at the rear of the police department), Mineola, 516-573-7620. **Hours**: Call for hours. **Fee**: Free. Features uniforms and equipment of the past, plus displays of K-9, mounted and other units as well as a 1925 station house, a 1973 communications bureau and vintage police vehicles.

Mitchel Athletic Complex, off Charles Lindbergh Boulevard (Meadowbrook Parkway Exit M4), Uniondale, 516-572-0400. This 47-acre tract has a 10,000-seat stadium for football, soccer and track that's home to many teams and events.

For a sampling of nearby restaurants, see Garden City-Uniondale on Page 284.

EAST HAMPTON

Where The Famous Are Never Far Away

O
K, SO NOBODY heads to the **Hamptons** for the history. But visitors who add museum-hopping to the usual whirl of sunning, shopping, dishing, noshing and star-gazing are guaranteed to encounter more celebrities (though granted, some may have been dead a century or two).

Luminaries have shaped the South Fork's resort villages for generations and their often deliciously gossipy legacy pervades the area — from the 18th and 19th century landmarks along **East Hampton's** broad, elm-shaded Main Street to the 20th century art colony shrines of rural **Springs** and the oceanfront trophy houses where Old Money meets New Media.

You won't find maps pinpointing current celebs' homes — this still isn't Hollywood, after all. But the haunts of past notables are more or less in plain sight, if sometimes open for limited hours or only by advance reservation. Such restrictions, however, are imposed to make your visit more enjoyable.

One of East Hampton's early Renaissance men (and theatrical superstars) was **John Howard Payne,** America's first actor to play Hamlet and to appear in Europe. He also was a diplomat, a poet and a writer of comedy, tragedy, melodrama and history. But he is immortalized for one opera lyric that he penned while homesick in Paris, supposedly dreaming of an East Hampton cottage he'd visited in his youth. That house is

Shops on Main Street in East Hampton

now named for his famous song: **"Home Sweet Home."**

There's no proof the globe-trotting Payne or any family member ever lived there (one theory is that 19th century neighbors who had a key concocted the tale after he became famous so visitors would pay them for a tour). But the myth lives on to a degree at the antiques-filled house at 14 James Lane, which has a lovely garden and a backyard windmill (though not in perfect working order like **Hook Mill,** a landmark on the green at the other end of the village). Home Sweet Home director **Hugh King** wears white gloves to handle the objets d'art and brims with intriguing esoterica — you have to pity bachelor Payne's unrequited "menage a nothing" (he was smitten with poet Percy Bysshe Shelley's widow, who loved writer Washington Irving).

As well as showing the house, the witty, knowledgeable King also leads village walking tours — including one of **South End Cemetery**, on the Town Pond green between James Lane and Main Street. Its weathered tombstones chronicle several centuries' passings of national significance: 1920s bons vivants Gerald and Sarah Murphy, whose soirees here and abroad included everyone from Pablo Picasso to F. Scott Fitzgerald; 19th century land-

Newsday Photo / Bill Davis

Clinton Academy, on Main Street, was the state's first chartered school.

Newsday Photo / Dick Kraus

The Hook Mill on North Main Street, built in 1806

scape painter **Thomas Moran,** a founder of the town's exclusive **Maidstone Club;** U.S. President John Tyler's son John, born to wife Julia of East Hampton's socially prominent Gardiner family (Tyler's summer White House adjoins Moran's former home).

Several buildings on this western end of Main Street are operated by the **East Hampton Historical Society** as museums: the 1740 **Osborn-Jackson House** (where several rooms are open as a decorative arts gallery), 1784 **Clinton Academy** (New York's first prep school), and the 1731 **Town House** (a playhouse-sized one-room school that once also served as the town hall). The present Village Hall across Main occupies the **Lyman Beecher House,** where the preacher father of writ-

AT A GLANCE

East Hampton Chamber of Commerce, 79A Main St. (down a narrow brick footpath) in the business district, 631-324-0362, www.easthamptonchamber.com. **East Hampton Historical Society, Osborn-Jackson House,** 101 Main St., 631-324-6850, www.hamptons web.com. Village parking is free, but most centrally located street spots have a one-hour limit, parking lots two hours; the small 24-hour area in the Lumber Lane lot fills by early morning in high season. Most beach parking is only with a resident permit on summer weekends. Partially wheelchair accessible. Child appropriate.

Did You Know? East Hampton was once part of Connecticut; it was settled by colonists from there who called it Maidstone for a town in their native England.

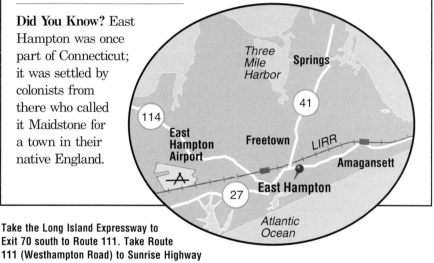

Take the Long Island Expressway to Exit 70 south to Route 111. Take Route 111 (Westhampton Road) to Sunrise Highway (Route 27). Go east on Route 27 for about 25 miles to East Hampton.

er Harriet Beecher Stowe began his family (it was after he'd relocated to Ohio, however, that daughter Harriet was inspired to write her antislavery novel, "Uncle Tom's Cabin"). Nearby **Guild Hall** has been a venue for art and theater greats and near-greats since 1931.

If you think East Hampton's business strip is nothing but designer shops, by the way, stop by the **Ladies Village Improvement Society** resale store — stocked with affordable books, clothes and household items. It's worth a look both for the building (another imposing Main Street mansion) and for the possibility that your bargain treasure might be a celebrity cast-off. Can a day's outing get any better than that? ◆

A Crucible of Expressionism

SOON AFTER **Jackson Pollock** and **Lee Krasner** bought a dilapidated farm there in 1945, the hamlet of **Springs**, just minutes from sophisticated East Hampton, changed from a baymen's backwater to an artists' colony.

The **Pollock-Krasner House and Study Center** stands as a shrine to those heady days in this rural crossroads — still readily identifiable only by a Presbyterian church, general store, tiny library and exhibit hall.

Tours of the Pollock property (by appointment only) provide a fascinating glimpse into the life of the ultimate starving, tortured artist — who fixed up the farmstead with a loan from his dealer-patron **Peggy Guggenheim** and often bartered paintings for groceries.

Visits start under two now large mimosa trees Pollock planted, where you sit on a pile of boulders he assembled as a free-form sculpture. Any paint speckles you might see on the rocks aren't his, though; they're the enthusiastic splashings of young art students dubbed **the Little Drippers.** But the barn-studio bears evidence of both artists' work — marks on the walls from Krasner's brush strokes, splatters covering the floor testimony to the unorthodox technique Pollock called "pouring" (he reportedly hated the critic who gave him the glib nickname **Jack the Dripper**). After donning foam slippers, visitors can walk over and around spatter patterns of some of his famous paintings, literally in his footprints.

Krasner had hoped the isolated peninsula would protect her husband from his wild city cronies, but the woods and salt marshes were never able to keep his demons fully at bay. He died on Aug. 11, 1956, when he lost control of his car just down the road during a drunken ride that injured his new girlfriend and killed a friend of hers (the crash scene in the 2000 film "Pollock" was staged in New Hampshire, because East Hampton was tired of movie shoots tying up traffic). Krasner, who was in Europe at the time of the accident, stayed in Springs until her death in 1984; the house contains mainly her possessions, plus a reference library.

Their graves — along with those of several dozen other artists and

Newsday Photo / Ari Mintz

Tools and paint cans remain in the converted barn that Jackson Pollock and his wife, Lee Krasner, used as a studio at their home in Springs.

writers — are in nearby **Green River Cemetery** on Accabonac Road. Both are marked by natural boulders — Krasner's smaller, of course, and located at Pollock's feet, where she had spent her life. ◆

Pollock-Krasner House and Study Center, 830 Fireplace Rd., Springs, 631-324-4929. Hours: Tours by appointment 11 a.m. to 4 p.m. Thursday to Saturday, May to October. Fee: $5.

Newsday Photo / Dick Kraus

At the East Hampton Town Marine Museum, a glass case holds a model of an inshore fishing boat, circa 1920, like those once used on Montauk's Fort Pond Bay. On the floor, a boat used for netting Atlantic menhaden.

WHILE YOU'RE THERE

East Hampton Town Marine Museum, Bluff Road, Amagansett, 631-267-6544. **Hours**: 10 a.m. to 5 p.m. daily in summer, weekends spring and fall. **Fee**: $4 adults, $2 seniors and younger than age 13. The museum, situated on a bluff overlooking the Nature Conservancy's largest double-dunes tract in the nation, depicts the history of whaling and fishing on the East End via dioramas and artifacts inside and on the grounds.

Amagansett Historical Association, Montauk Highway at Windmill Lane, Amagansett, 631-267-3020. **Hours**: 9 a.m. to 1 p.m. Friday to Sunday late June to early September. **Fee**: $3, $1 younger than age 12. The **Miss Amelia Cottage Museum**, built in 1725 by village founding merchant brothers Abraham and Jacob Schellinger, houses changing exhibits of Amagansett life from colonial times to the early 20th century. Furnished with local Dominy furniture, including clocks and 18th and 19th century household artifacts and tools. Also

Newsday Photo / Ken Spencer

Russian sage in the garden at LongHouse Reserve, home of Jack Lenor Larsen, in East Hampton. LongHouse aims to create landscapes as an art form and to demonstrate potential plantings in the climate of the East End.

there is the **Roy K. Lester Carriage Museum** and **Jackson Carriage Barn** housing 28 horse-drawn vehicles.

LongHouse Reserve, 133 Hands Creek Rd., East Hampton, 631-329-3568. **Hours:** Open 2 to 5 p.m. every Wednesday and Saturday from late April to early September. **Fee:** $10, younger than age 12 free with an adult. The reserve has a gallery and 16-acre sculpture-filled gardens founded by textile designer and collector Jack Lenor Larsen in 1991.

Madoo Conservancy, 618 Main St., Sagaponack, 631-537-8200. **Hours:** 1 to 5 p.m., Wednesday and Saturday May to September. **Fee:** $10, no one younger than age 6. The 2-acre gardens meld the winter and summer houses and studios of artist and plantsman Robert Dash. The gardens are studded with 18th century shingled structures built of shipwreck timber. There's also a vegetable garden, 120-foot rose walk and a Chinese bridge.

For a sampling of East Hampton restaurants, see Page 282.

EISENHOWER PARK

A Big Park That Goes The Distance

MARATHONS (and marathoners' Olympic dreams) start and finish there. The next Tiger Woods or Venus Williams could be practicing on an adjacent golf green or tennis court when you're playing there. And on any given day, thousands of walkers, joggers, skaters, golfers, bicyclists, ballplayers, scooter riders, kite fliers, stroller pushers, model boaters, lawn bowlers, concertgoers and Canada geese amicably coexist there (except sometimes the geese).

Where? Precisely in the middle of Nassau County at 930-acre **Eisenhower Park** — larger than Manhattan's Central Park by 87 acres and boasting many features that city folks probably wish their park had. Such as **three 18-hole golf links**, two miniature-golf courses with 18 holes each, a golf driving range, batting cages, reservable picnic areas and the **largest pool** in the Northern Hemisphere (at a year-round state-of-the-art swimming-diving center).

The only spoiler in this barrel of fun is that all its bounty is intended for the enjoyment of only Nassau County residents and their guests — with the notable exception of the aquatic center, which was built primarily with state funds. And even there, holders of the requisite proof of residence — a $15 **Nassau County Leisure Passport** — generally pay about

Skaters and strollers on one of many paths at Eisenhower Park

half the fees charged nonpass holders to use aquatic center facilities.

Eisenhower's wide range of activities easily make it Nassau County's most popular park. Only one main road (Park Boulevard) winds across it, though, and while there are numerous parking areas, you still have to walk a modest amount (but isn't that why you came?). The asphalt paths are in good condition, and although the park is more active than scenic, the fact that it comprises flat, open fields makes it easy going. Before you know it, you've covered a lot of ground: circled **Salisbury Lake** (favorite haunt of model boaters, as well as the hissy geese), listened to the gleeful laughter at the various playgrounds, checked on what's growing in the **American Dahlia Society Eastern Trial Garden**, watched a bridal party pause for photos outside Carltun on the Park restaurant and paused at various memorials: large monuments for war veterans and firefighters, smaller ones for causes and individuals (there's one in memory of children of the Holocaust, another honoring John F. Kennedy for being the first president who as a youth was a Boy Scout).

Free summer concerts are held Wednesday noons and numerous evenings (as many as 15 nationalities have been celebrated on ethnic nights, and an international festival is held each year during the first weekend in August). And year-round, countless sports leagues and other organizations meet in the park for athletic or recreational activities.

That fenced-off miniature village you might notice just inside the Stewart Avenue entrance is called **Safety Town**. It's used by the Nassau County Police Department to teach bicycle and pedestrian safety to third-graders, mainly in school groups, but in July and August it's open to all Nassau County children ages 7 to 9 weekdays from 12:30 to 2:30 p.m. A distant corner of the park also houses the county mounted police unit.

Newsday Photo / J. Michael Dombroski

Along with a driving range, Eisenhower Park offers three 18-hole golf courses.

In the 1920s, most of the present park land belonged to the posh Salisbury Country Club — which even had its own Long Island Rail Road siding. When the owner fell upon hard times and couldn't pay the taxes, the county (oh, the irony) acquired the property. In 1944, it established the **Nassau County Park at Salisbury**, which was rededicated as the Dwight D. Eisenhower Memorial Park in 1969, several months after the former president's death. On hand for the ceremony were two nonresident notables: Ike's grandson, Dwight D. Eisenhower II, and his wife, Julie Nixon Eisenhower. ◆

AT A GLANCE

Eisenhower Park, East Meadow (entrances at junctions of Stewart and Merrick avenues and of East Meadow Avenue and Hempstead Bethpage Turnpike). Call 516-572-0200 daytime Monday to Friday; 516-572-0252 after 6 p.m. to confirm evening shows on date of performance; 516-572-0223 for recorded weekly updates on all Nassau County-operated parks and museums. **Hours**: Dawn to dusk year-round (swimming-diving center and various lighted activities have later hours). **Fee**: Free, but some activities have a charge. Wheelchair accessibility. Child appropriate.

Eisenhower Park Swimming & Diving Center (in the park), 516-572-0501. **Hours**: Monday to Friday 9 a.m. to 10 p.m., Saturday and Sunday 9 a.m. to 7 p.m. (hours subject to change during special events). Extended morning hours (6 to 9 a.m.) for members Monday to Saturday. **Varying fees**. Wheelchair accessibility. Child appropriate.

Park and pool Web site: www.co.nassau.ny.us/parks.html. Call 516-572-0200 for information on Leisure Passports, which provide entry to all park facilities (and discounts on ones with a charge), available to Nassau County residents ages 13 and older (they're required of residents ages 10 and older for golf). They're good for three years (or for senior citizens and residents with disabilities, for as long as they live in Nassau).

Did You Know?
It's believed that English colonists named the land now occupied by Eisenhower Park after England's Salisbury Plain.

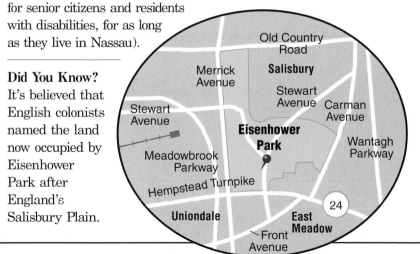

From either the Northern State Parkway or the Southern State Parkway, take the Meadowbrook Parkway to Exit M5 east (Route 24). Go east about one-half mile to the park entrance on the left.

Newsday Photo / J. Michael Dombroski

The $30-million Nassau County Aquatic Center opened in 1998.

Wet and Wonderful

THE MULTILEVEL **Eisenhower Park Swimming & Diving Center** has hosted superstar aquatic events since it opened with the 1998 Goodwill Games. In 1999, it was the site of the **Empire State Games** and in 2001, its world-class natatorium — equipped with what's billed as the largest pool in the northern hemisphere — featured the **NCAA Division I women's swimming and diving championships**.

It's not exclusively for Nassau County residents because it was built primarily with state funds (through the New York State Dormitory Authority).

In addition to its **segmented pool** — whose floor can be raised or lowered in one section, depending on whether groups of children or adults are using it — there's a **whirlpool** and a **sauna**. The building also contains fitness facilities and there's an outside terrace area with tables and chairs. ◆

WHILE YOU'RE THERE

African American Museum, 110 N. Franklin St., Hempstead, 516-572-0730. **Hours**: 6 to 9 p.m. Wednesday; 10 a.m. to 4:45 p.m. Thursday to Saturday; 1 to 4:45 p.m. Sunday. Some program fees. The museum deals with the history and contributions of Long Island's African-American community. It offers films, lectures and special programs year-round. Arts and crafts workshops are held in July and August.

Newsday Photo / Bill Davis

Kwanzaa celebration at the African American Museum in Hempstead

Hofstra University on Hempstead Turnpike in Hempstead, 516-463-6600, has a museum, arboretum and a regular calendar of cultural and sports events open to the public. Guided campus tours geared to prospective students can be arranged by calling 800-463-7872.

The entire 238-acre Hofstra campus is registered with the American Association of Botanical Gardens and Arboreta, and its more than 8,000 trees represent 425 species and varieties; a 2-acre plot has been developed as a distinctive bird sanctuary. Information and brochures about self-guided tours are available by calling 516-463-6815.

The **Hofstra Museum** has a permanent collection of more than 4,500 paintings, drawings, prints, sculptures, anthropological and decorative art objects. For information on exhibits and gallery hours, call 516-463-5672.

Nassau Veterans Memorial Coliseum, 1255 Hempstead Tpke., Uniondale, 516-794-9300, offers a year-round slate of sports events, concerts and exhibitions.

For a sampling of nearby restaurants, see Garden City-Uniondale on Page 284 and Westbury area on Page 296.

DISCOVER LONG ISLAND

A Fragile Resort Of Dunes And Dreams

ACH SUMMER without fail, a flotilla of some 3 million singles, families, loners, groupers, straights, gays, celebrities and homebodies descends on **Fire Island** — the stunning 32-mile-long sandbar that is New York's Key West (minus the palm trees and Hemingway haunts).

Through house-wrecking hurricanes (and parties), antediluvian restrictions (such as the outdoor eating bans that labeled popular Ocean Beach **"The Land of No"**) and zany traditions (notably the mascara-laden annual invasion of the more sedate of its two gay-oriented communities by denizens of the cheekier one), it has remained the ultimate offshore getaway.

Anchored in the Atlantic a few miles off Long Island's South Shore, it's accessible to the public by car only as far as parking lots at **Robert Moses State Park** at the western end and **Smith Point County Park** at the eastern tip. Thus day-trippers and homeowners alike must rely on private boats, water taxis and the frequent seasonal passenger ferries that serve about a dozen of its 17 major communities. A rutted sand trail dubbed the **Burma Road** runs along the island's spine, but there are no other streets — only wooden or cement walkways. Officials patrol in modified golf carts and dune buggies and, in wilderness areas, on horseback.

Since 1964, all of Fire Island except the six-mile stretch of state park (an

66

Passengers arrive at Fire Island after riding the ferry from Bay Shore.

easy stroll to Fire Island Lighthouse and the westernmost settlement of Kismet) has been part of **Fire Island National Seashore**. But by then, local opposition had saved the island from a major development planned by Robert Moses — perhaps his only bad idea, and only failure, as head of the Long Island State Parks Commission. (The Seashore also includes about two dozen Great South Bay isles plus the mainland estate of patriot William Floyd.)

The island's population swells from about 600 year-rounders to a summertime 25,000, who somehow squeeze (some only on alternate weekends) into 4,000 houses identified by cute names instead of numbers. Communities are as different as seashells but do have some things in common: Most are a mix of singles and families who love this fragile resort held together by dune grass and dreams — and who, partly because no one has to drive home, throw legendary cocktail parties.

Most favored by daytrippers are three nonresidential National Seashore destinations — **Sailors Haven, Watch Hill** and **Talisman / Barrett Beach** — plus the communities of **Ocean Beach, Ocean Bay Park** and gay-but-egalitarian **Cherry Grove**. In the late-1800s, the island was dotted with huge hotels, but today it has just a handful of small hotels and guest houses — in these places and upscale gay **Fire Island Pines**. The park service's latest concessionnaire contract also allows approved modest accommodations at

Newsday Photos / Michael E. Ach

A family enjoys a September day at Ocean Beach.

Wagons await their owners' burdens at Ocean Beach, where car traffic is banned.

its areas in the future.

Besides swimming and sunning, what's there to do? More than you might think. At Sailors Haven, you can explore the island's unusual natural feature: the rare **Sunken Forest,** with roots going back to the American Revolution. Ocean Beach, the Pines and the Grove have small shopping districts, but a favored daytime pastime in almost any community is strolling to admire the rose-covered beach cottages and Malibu mansions that cling precariously to the shifting sand (and help stabilize it). The only off-limits village is the island's first, **Point O' Woods.** Founded in 1894 as a Chautauqua Assembly, it has its own ferry and hides amid trees behind a chain-link fence.

In addition to occasional concerts, art

AT A GLANCE

Fire Island Tourism Bureau, 631-563-8448 (seasonal). **Fire Island National Seashore**, 631-289-4810. Also helpful: www.fireisland.com, www.fireislandbeaches.com, www.barrierbeaches.com, www.obvillage .com. A detailed Fire Island Guide ($4.95) is published by Ocean Beach's summer weekly, Fire Island News. The island's other paper, Fire Island Tide, is published biweekly in Davis Park. Sayville Ferry Service, 631-589-8980; Davis Park Ferry Co. in Patchogue, 631-475-1665; Fire Island Ferries in Bay Shore, 631-665-3600.

Did You Know? Theories abound, but Fire Island's name probably stems from nighttime blazes that guided early whalers ashore (and lured pirates' prey).

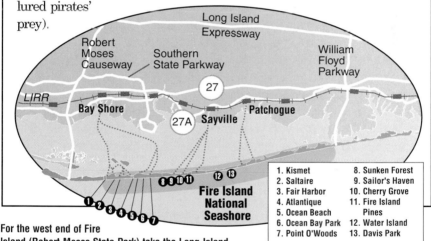

Long Island Expressway

Robert Moses Causeway

Southern State Parkway

William Floyd Parkway

LIRR

Bay Shore

27

27A Sayville

Patchogue

Fire Island National Seashore

1. Kismet
2. Saltaire
3. Fair Harbor
4. Atlantique
5. Ocean Beach
6. Ocean Bay Park
7. Point O'Woods
8. Sunken Forest
9. Sailor's Haven
10. Cherry Grove
11. Fire Island Pines
12. Water Island
13. Davis Park

For the west end of Fire Island (Robert Moses State Park) take the Long Island Expressway to Exit 53 to the Sagtikos State Parkway. Drive south on Sagtikos to the Southern State Parkway and follow signs for Robert Moses Causeway south. Take the causeway into Robert Moses State Park. For the east end of Fire Island (Smith Point County Park) take the Long Island Expressway to Exit 68 to the William Floyd Parkway south to Smith Point County Park.

shows, kids' events and social and civic fund-raisers, there's theater in the Grove and the Pines (not all gay-themed), movies in Ocean Beach and a pulsing bar scene in these places plus Ocean Bay Park.

Fire Island has always had its celebrity habitues, but unlike the Hamptons set, most truly seek anonymity (oh, all right, just one long-timer to keep an eye out for: producer, director, actor and writer Mel Brooks). ◆

A Forest Primeval

E VEN MIGRATING BIRDS may do a double take when they spy the dense 40-acre woodland on windswept Fire Island.

The rare **Sunken Forest** developed at **Sailors Haven** because a double line of oceanfront dunes formed there over the centuries, providing extra protection against the gales. First grass, then hardy shrubs and scrub pines, then hardwood trees took root until now there's an old- growth forest snarled with briars and vines. Its bogs support vegetation such as ferns and swamp maple more common to freshwater wetlands, but ground cover is rare due to both the ever-chomping deer and the sun-blocking overhead canopy. (Predominant giants include American holly and sassafras, the latter notable for producing leaves of several different shapes on a single tree.) The entire forest is pruned even with the top of the primary dune by the salt spray — which ironically also provides nutrients.

A 1.5-mile round-trip boardwalk weaves across this forest primeval, a bay-side spur threads through head high marsh grass. You can explore on your own or take a ranger-led walk. But you must stay on the boards or the alternate beach trail — to protect the environment and to lessen your chance of a brush with poison ivy or deer ticks. Winged wildlife, depending on season, can range from butterflies to falcons. ◆

Fire Island National Seashore, 631-289-4810; ferries to its three park facilities operate from May to October.

Sailors Haven offers swimming in July and August (lifeguards, bathhouses and showers); a visitor center with fish tanks and other exhibits; a snack bar and gift shop (ferry from Sayville).

Talisman / Barrett Beach is an undeveloped area reached by private boat or walking east about a mile from Fire Island Pines.

The seven-mile Otis Pike Fire Island High Dune Wilderness Area, about three miles farther east, is bounded on the east by the Fire Island Wilderness Visitor Center (reachable via Smith Point causeway), with exhibits plus information about backcountry camping, hiking, fishing and hunting. The area's western anchors are the community of Davis Park and the national seashore's adjacent Watch Hill Visitor Center (both reached by ferries from Patchogue). Watch Hill offers a nature trail, a large marina and a bare-bones campground. (Reservations required for camping. Call 631-597-6633 or 631-289-4810; also see http://www.nps.gov/fiis.

Newsday Photo / Michael E. Ach

A path through Fire Island's Sunken Forest

Savvy campers favor spring or fall when there are no mosquitoes.)

For information on Robert Moses State Park and the Fire Island Lighthouse, see Pages 192 to 197. For information on Smith Point County Park, see Page 279.

For a sampling of nearby restaurants, see Bayport-Patchogue area on Page 280 and Sayville on Page 293.

FREEPORT

Teeming Nautical Mile Is a Great Catch

L IKE A BUSTLING Mediterranean fishing village, Freeport's Woodcleft Canal waterfront is awake before dawn. Seagulls swoop and squawk as the returning commercial fleet delivers the night's catch to dockside markets and restaurants. Then by 8 a.m., amateur anglers arrive with their lucky hats and Dramamine, and the barnstorming gulls quickly abandon the empty trawlers for the parade of charter boats heading for a half-day of hopeful casting in the sheltered bays or open ocean beyond.

A couple of hours later, shops along **Woodcleft Avenue** start flipping over their window signs from Closed to Open, and chairs and tables sprout at sidewalk cafes. By noon, two overgrown yachts have shoved off on the first of their two daily casino cruises and the first of countless pounds of steamers are being downed.

So begins another day on **Long Island's Nautical Mile**, a longtime landmark for seafood lovers that in recent years has been revitalized by renovations and enhancements: new shops and restaurants, a waterfront esplanade, a "Scenic Pier," flower boxes, antique light fixtures and a customized branch of Manhattan's South Street Seaport Museum (here dubbed the **"Seaport at Freeport"**). In the evening, party boats set sail for dinner cruises and private functions while ashore, music beckons strollers into crowded

At Capt. Ben's Fish Dock on Woodcleft Avenue, diners can enjoy the Bracco Clam & Oyster Bar, above, in the summer. Capt. Ben's is open all year.

outdoor bars exuding such a jovial islands mood that you almost forget you're not in the village's namesake vacation paradise in the Bahamas. Freeport has got its groove back.

The Nautical Mile — extending the entire length of Woodcleft Avenue, from Front Street to the Scenic Pier — is an easy stroll that passes in what seems more like a nautical minute because there's so much to see. The handful of gift shops grew to two handfuls in 2001 with the addition of a small, beachy-looking complex called **Crow's Nest Cove**. No match for a mega-mall, to be sure, but Woodcleft's craft show quality wares are well-priced and you're not likely to leave empty-handed. Both

A prize red snapper at Two Cousins Fish Market.

Newsday Photo / J. Michael Dombroski

A Hans Gabali mural on Woodcleft Avenue

sides of the avenue are crammed with marine supply businesses — one showroom bills itself a "yacht factory outlet clearance center." There are boats everywhere, from the multi-tier outdoor parking lots for them to, of course, the water.

As you meander along, be sure you don't miss Woodcleft's two grand wall murals by Northport's **Hans Gabali** (one is at the corner of Hamilton Street, the other between Hamilton and Suffolk).

You can pause at any (or all) of several ice cream parlors for a hot fudge sundae. Or amble along with a cone to the **Scenic Pier** — actually a wide covered deck at the bay end of the street. Then sit a while, watching the water and the palette of artists usually working at easels to immortalize the passing boats and surrounding salt marshes. The village is weighing further developments on the adjacent open acreage. Among the possibilities: a Jones Beach water shuttle.

Ready for a more substantial meal? You can choose from close to two dozen eateries, from casual clam bars to more upscale restaurants — most

AT A GLANCE

Freeport Chamber of Commerce, 300 Woodcleft Ave.,
516-223-8840, www.freeportny.com. There are free parking lots
between Woodcleft and South Ocean avenues. Wheelchair accessible.
Child appropriate.

Did You Know?
Freeport got its
name from colo-
nial-era sea
captains, who
could unload
goods there
without paying
the usual cus-
toms fees charged
by other ports.

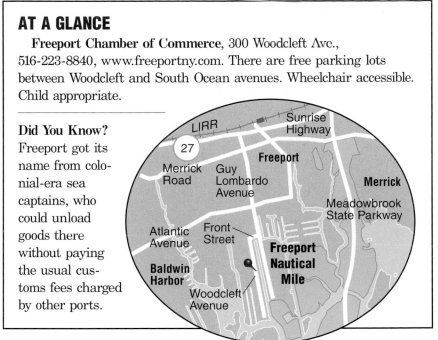

Take the Meadowbrook Parkway south from
the Northern State Parkway to Sunrise Highway
(Route 27). Go west to Guy Lombardo Avenue in Freeport and turn left; proceed past
John B. Randall Park, then right onto Front Street. At next corner go left onto Woodcleft Avenue
— the beginning of Freeport's Nautical Mile.

with indoor-outdoor dining and water views. Or you can purchase fish
fresh from the boat to cook at home.

Fishing tournaments, a seafood festival, plus the annual August visit
from the U.S. Naval Academy Sailing Training Squadron keep things lively
through the summer. There's also a fall nautical festival.

Celebrities always have gravitated to Long Island's South Shore, and
Freeport's most famous — bandleader **Guy Lombardo** — moved to the
Woodcleft Canal's residential eastern shore in 1940. For many years his
name graced a restaurant at the end of the main road there — now called
Guy Lombardo Avenue. His house and the restaurant are gone, but a marina named for him is site of an annual in-water boat show at the beginning
of summer. And old-timers who fondly remember Lombardo as Mr. Freeport, muse, "If he could see it now." ◆

A sculpted sea captain at the entrance to the "Seaport at Freeport"

Tales and Treasures

S INCE THE **"Seaport at Freeport"** arrived on the scene several years ago, memorabilia from the South Shore's maritime past has flowed in like a full-moon tide. And when the museum opens its wall-sized side doors during village festivals, its displays even more visibly meld into the ebb and flow of life along the **Nautical Mile**.

"We're still a living, working street, so we concentrate as much as possible on the local area," said education director Helen Watkin, whose soft English accent seems right at home in this multicultural village where a common bond seems to be saltwater coursing through the veins.

Ship models, handmade fishing lures, buoys, knots, lanterns and actual boats are hanging, leaning and otherwise displayed everywhere in the comfortably cluttered museum (formally the **South Street Seaport Museum Long Island Marine Education Center**). There's a rare racing hydroplane that belonged to Freeporter Guy Lombardo, a handmade sailing canoe donated by a Baldwin woman who was courted by her husband in it, an 1880 duckboat that once plied the nearby bays, a Jones Beach lifeboat, an old

---

WHILE YOU'RE THERE

Freeport Historical Museum, 350 S. Main St., Freeport, 516-623-9632. **Hours**: Open Sundays 2 to 5 p.m., April through December. **Fee**: None, but donations appreciated. This museum features furnishings and clothing of bygone eras exhibited in a Civil War-era bayman's cottage.

Oceanside Marine Nature Study Area, 500 Slice Dr., Oceanside, 516-766-1580. Operated by the Town of Hempstead. **Hours**: Open 9 a.m. to 5 p.m. Tuesday to Saturday year-round except holidays. **Fee**: None. Features 52 acres (mainly salt marsh) with trails, indoor and outdoor displays (mostly local marine specimens), bird-watching; guided tours can be arranged in advance.

Cow Meadow Park Preserve, South Main St., Freeport, 516-571-8685. Operated by Nassau County. **Hours**: Dawn to dusk (later for some activities). This 150-acre preserve has a quarter-mile hiking trail through bayberry thicket along a marsh, and 156 species of nesting, migrating and wintering birds; small bird-watching tower.

For a sampling of Freeport restaurants, see Page 283.

Coast Guard copper-bottom "rollover" surf boat, a replica rumrunner (During Prohibition, **Freeport Point Shipyard** built these speedy craft as well as the Coast Guard boats that chased them) and the piece de resistance: a dugout canoe with the girth of an elephant, found in the late 1990s in the mud of Milburn Creek between Baldwin and Freeport (it hasn't yet been authenticated, but is believed to be the work of **American Indians** — which could put its origin in the early 1700s). Posters and dioramas depict earlier seafaring times, and a seasoned fishing boat, also indoors, welcomes aboard the regular onslaught of inquisitive young visitors.

Various art projects are usually interspersed among the artifacts, as well. Watkin, herself an artist, has been working on an indoor mural with Hans Gabali and some school students. Plus, you never know what new treasure the next tide might wash in. ◆

South Street Seaport Museum Long Island Marine Education Center, 202 Woodcleft Ave., 516-771-0399. Hours: Tuesday to Friday 11 a.m. to 4 p.m., Saturday to Sunday 1 to 5 p.m. year-round. Fee: $2 adults; family activities 2 p.m. Sundays.

GARVIES POINT
MUSEUM & PRESERVE

A Glimpse
Of the Island's
Distant Past

MANY LONG ISLAND communities boast a few centuries-old houses, their character and chronology documented by early residents' diaries. But where can you get a glimpse of life here 5,000 years ago, before the advent of tell-all books?

You might start with the excavation model of a prehistoric North Shore Indian campsite at **Garvies Point Museum & Preserve** in Glen Cove.

Matinecock Indians occupied a sheltered portion of what's now the museum property as early as 2500 BC. It was a small seasonal site with fewer than 50 residents, who were primarily hunters and gatherers of wild foods such as shellfish. Museum exhibits also depict Indian life before the diseases and opportunistic economy of European settlers in the 17th century brought a rapid decline to an ancient culture (though after a short walk along the eastern shore of **Hempstead Harbor**, you can catch a Pequot-operated ferry linking Glen Cove with that Connecticut tribe's latter-day resort casino).

Nassau County's South Shore, too, was once the home of a thriving Indian community, and the museum displays about 14 dozen arrowheads

An inside look at the Garvies Point Museum in Glen Cove

discovered in Massapequa Lake when it was drained for a construction project. Along with the **"Massapequa Lake Blade Cache"** are objects recovered from a dig at a 17th century Indian stronghold called Fort Massapeag. These items include part of a palisade post and a mortar — extremely rare finds for Long Island, whose acidic soil typically destroys wood artifacts. The fort was used by the **Marsapequa Indians** for protection against the English and Dutch in times of danger.

You can reach back much farther in time than 4,500 years — 570 million years, to be exact — via museum exhibits on Long Island geology, which include dinosaur and other fossils. Inside the museum, you can trace time lines showing how changes in climate and sea level have caused our landscape to evolve in the 20,000 years since the most recent advance of the continental ice sheet (some rocks outside, which were swept far from their point of origin and deposited here, bear the scars of their glacial journey).

The preserve offers opportunities to explore many other geologic features that are typical of Long Island but not found together elsewhere in such a concentrated area. Four miles of trails wind through 62 acres of glacial moraine covered by meadows, forests, and thickets of shrubs and vines that attract more than 140 species of birds.

On one section of beach, erosion of the high cliffs that form a backdrop reveals veins of age-old multicolored clays oozing out of the slumping hillside. Many well-preserved imprints of plants dating back hundreds of millions of years have been found in the red shale and sandstone along that stretch of shore. In modern times, the preserve was part of the estate of **Thomas Garvie**, a physician who immigrated to Glen Cove from Scotland in the early 19th century. The trail map also will lead you to his family cemetery, where several names and dates are inscribed on an 11-foot-high boulder. There also are occasional guided nature walks.

Don't worry, you can get your hands as well as your feet dirty there, too. The museum, operated by the Nassau County Department of Recreation & Parks, offers a year-round schedule of educational programs on local geology, Long Island Indian culture and archeology. An annual **Native American Feast** the weekend before Thanksgiving includes an array of hands-on activities, craft demonstrations, food displays and samples. **Garvies Point Day** (the first Saturday of August) features activities based on a different theme each year. And while March is usually Geology Month, you can bring puzzling stones, arrowheads and other treasures anytime to periodic workshops on identifying fossils, rocks and minerals. ◆

Newsday Photo / Ken Spencer

This exhibit at the Garvies Point Museum focuses on activities of American Indians throughout the year.

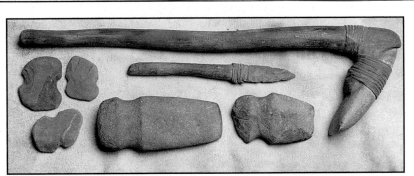

Newsday Photo / Bill Davis

At the Garvies Point Museum, woodland artifacts used by early inhabitants of Long Island

AT A GLANCE

Garvies Point Museum & Preserve, 50 Barry Dr., Glen Cove; 516-571-8010. **Hours**: Open Tuesday to Sunday 10 a.m. to 4 p.m. except winter holidays. **Fee**: adults $2, ages 5 to 12 $1. Partial wheelchair accessibility. Child appropriate.

Did You Know? New York's state fossil is the rare eurypterid, or sea scorpion — whose relatives include horseshoe crabs, spiders and lobsters.

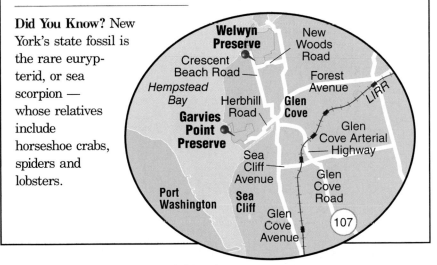

Take the Long Island Expressway to Exit 39 north (Glen Cove Road). Stay on Glen Cove Road north until you see Glen Cove Arterial Highway on the left. Take the arterial and continue to the last traffic light at the end of the road, facing the Glen Cove firehouse. Turn right and follow directional signals to the museum and preserve.

To Never Forget

A 204-ACRE Glen Cove estate where the international set partied during the Roaring '20s today serves a deep humanitarian purpose.

The centerpiece 1910 Georgian mansion on what is now the tranquil **Welwyn Preserve** has become the **Holocaust Memorial & Educational Center** of Nassau County — established to honor victims of the Nazis and their collaborators, as well as to work toward eliminating prejudice and hate crimes through enlightenment.

Four marked nature trails wind through old-growth woods, overgrown meadows, and relics of the days when the estate was the home of Harold Irving Pratt and his wife, Harriet Barnes Pratt. It was one of six Pratt homes on a 1,100-acre tract oil baron Charles Pratt divided among his sons in the early 1900s. The estate became property of Nassau County in 1974; the Holocaust Center moved there about 20 years later. The philanthropic Pratts, who also founded Brooklyn's Pratt Institute, would certainly approve.

The center offers age-appropriate workshops, lectures, performances, and other programs for children and adults. Permanent exhibits include posters demonstrating the chronology of the Holocaust as well as video testimony surrounding photographs of the World War II period. Changing displays feature related works by local artists and schoolchildren.

Newsday Photos / Ken Spencer

A display at the Holocaust Memorial & Education Center of Nassau County at Welwyn Preserve in Glen Cove

A major current project is restoring to its former glory one of the estate's acclaimed formal gardens — designed by Olmstead Brothers, the firm founded by the famous landscape architect **Frederick Law Olmstead**. In its heyday voted the most beautiful in America, the revitalized garden will become a memorial to children of all faiths who were Holocaust victims.

Children also are con-

WHILE YOU'RE THERE

Other sites to visit in the Garvies Point-Welwyn area include:
Planting Fields Arboretum State Historic Park (see Page 178)
John P. Humes Japanese Stroll Garden (see Page 108)
Sea Cliff Village Museum (see Page 228)
Bailey Arboretum (see Page 113)
For a sampling of Glen Cove restaurants, see Page 284.

The sculpture "Freedom" commemorates the victims of the Holocaust.

sidered key to the center's mission. School groups are told: "You are the future. You will shape this country. Upon your decisions will rest our country's morality . . . Remember, people who lived during the Holocaust fell into four classes: victims, bystanders, perpetrators and rescuers. Promise that you will never allow yourself to be any of the first three, but only the last." ◆

Holocaust Memorial & Educational Center of Nassau County at Welwyn Preserve, 100 Crescent Beach Rd., Glen Cove, 516-571-8040 or on the Internet at www.holocaust-nassau.org. Hours: Open 9:30 a.m. to 4:30 p.m. Monday to Friday, 11 a.m. to 4 p.m. Sunday. Fee: Free.

GOLD COAST MANSIONS

Glittering Domains Of LI's Royalty

THEY WERE THE ultimate gated communities. The several hundred French chateaux and English castles built by America's movers and shakers a century ago on Long Island's North Shore created a sort of millionaires' row dubbed the **Gold Coast**.

For several glittering decades, from the Gay '90s through the Roaring '20s, these opulent estates were home to prominent families with names such as Vanderbilt, Astor, Chrysler, Morgan and Tiffany. Unlike the sometimes even grander seasonal "cottages" of Newport, R.I., and other resort areas, the mansions of Nassau and western Suffolk counties offered a rare perk: country living within an easy commute of America's financial capital by train or car (and the occasional 200-foot private yacht).

Many Gold Coast mansions reverberated with the decadent Jazz Age parties **F. Scott Fitzgerald** depicted in "The Great Gatsby." But these were philanthropists and Renaissance men more than sybarites. If they spared no expense on their homes, it was because they planned to bequeath them as public museums or parks.

What killed the legendary estates, with their polo fields and their children's playhouses the size of the modest tract homes that replaced some

The house and grounds of Falaise, at the Sands Point Preserve

mansions? A triple whammy dealt by the advent of income and property taxes, the financial losses of the **Great Depression**, and changes in the U.S. economic structure that made domestic service a less attractive job for the legions of workers needed to keep this way of life humming.

About 200 of these Gatsby-esque mansions still exist here — most now private enclaves used as country clubs, schools and religious institutions

Newsday Photo / Bill Davis

Hempstead House dressed up for a holiday celebration

that may be open under some cir-
cumstances or by special ar-
rangement. But about a dozen
along an approximately 25-mile
stretch between **Kings Point** and
Centerport are regularly open to
the public — filled with customized architectural details and
often treasures that their owners
collected from around the world.
While you couldn't possibly tour
all in one day, you can easily
work them into a variety of
North Shore excursions.

In an effort to provide a focus
for such visits, Suffolk County recently reopened the Museum of
Long Island's Gold Coast in Huntington — naturally in an old
mansion, most recently known

Newsday Photo / Michael E. Ach

*A staircase at the Nassau County
Museum of Art in Roslyn Harbor*

MANSIONS AT A GLANCE

1. Wiley Hall at the U.S. Merchant Marine Academy, 516-773-5000.

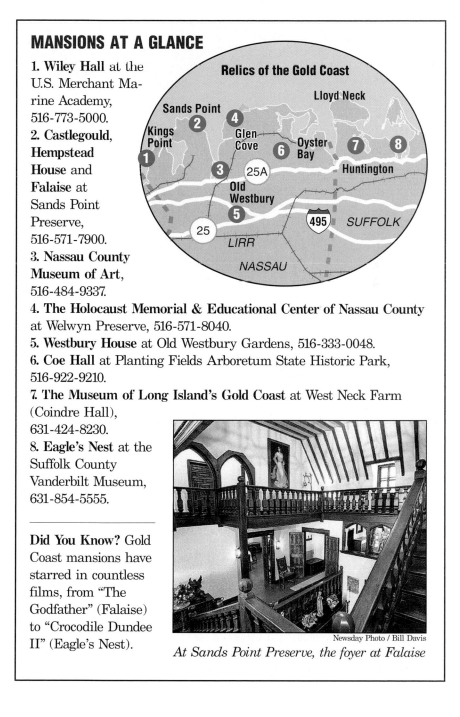

Relics of the Gold Coast

Sands Point
Kings Point
2 Glen Cove **4** Lloyd Neck
1 **3** Old Westbury
25A
5 25 LIRR
6 Oyster Bay **7** **8** Huntington
495 SUFFOLK
NASSAU

2. Castlegould, Hempstead House and **Falaise** at Sands Point Preserve, 516-571-7900.

3. Nassau County Museum of Art, 516-484-9337.

4. The Holocaust Memorial & Educational Center of Nassau County at Welwyn Preserve, 516-571-8040.

5. Westbury House at Old Westbury Gardens, 516-333-0048.

6. Coe Hall at Planting Fields Arboretum State Historic Park, 516-922-9210.

7. The Museum of Long Island's Gold Coast at West Neck Farm (Coindre Hall), 631-424-8230.

8. Eagle's Nest at the Suffolk County Vanderbilt Museum, 631-854-5555.

Did You Know? Gold Coast mansions have starred in countless films, from "The Godfather" (Falaise) to "Crocodile Dundee II" (Eagle's Nest).

Newsday Photo / Bill Davis

At Sands Point Preserve, the foyer at Falaise

Newsday Photos / Ken Spencer

Inside Wylie Hall at the U.S. Merchant Marine Academy at Kings Point

as **Coindre Hall**. The original estate name, West Neck Farm, will be re-instated as the property is slowly restored.

Other Gold Coast mansions that can be visited, starting with the west-ernmost, include those that follow. Hours vary by day and season, and some can be visited only on group tours.

• **Wiley Hall** at the U.S. Merchant Marine Academy at Kings Point is the former home of auto magnate Walter P. Chrysler.

• Sands Point Preserve, former estate of railroad tycoon Jay Gould's son Howard, has three mansions: **Castlegould** (housing temporary exhibits), **Hempstead House** (displaying a collection of Wedgwood china) and **Falaise** (filled with mementos of philanthropist Harry F. Guggenheim, who built it after his parents purchased the Gould estate).

Hempstead House, at Sands Point, is open for tours.

- **Nassau County Museum of Art** in Roslyn Harbor offers indoor and outdoor art on a former Frick family estate.
- **The Holocaust Memorial & Educational Center of Nassau County** in Glen Cove's Welwyn Preserve occupies one of six homes that once graced a family compound created by oil baron Charles Pratt.
- **Westbury House** at Old Westbury Gardens is the former home of steel-fortune heir John "Jay" Phipps and shipping heiress Margarita "Dita" Grace Phipps.
- **Coe Hall** at Planting Fields Arboretum State Historic Park in Oyster Bay was the residence of insurance executive William Robertson Coe.
- **Eagle's Nest** at the Suffolk County Vanderbilt Museum in Centerport was home to William K. Vanderbilt II. ◆

New Life for Past Glory

T HAT CLUSTER of bare lightbulbs hanging over the spiral teakwood staircase in the **Museum of Long Island's Gold Coast** is actually a symbol of the once-grand mansion's heyday, not its fall from grace.

When the house was built in 1912, electricity was new, too, so anyone who could afford it showed it off (the innovative home, now a national landmark, also had central heating and vacuuming). Though few furnishings remain from those heady days, tours fill in the gaps and a photo exhibit at the still-evolving museum details the genre of North Shore country houses. And the fate of this one estate, called **West Neck Farm**, also exemplifies the rise and fall of the Gold Coast.

The farm, on the Huntington-Lloyd Harbor border, was once the 54-acre waterfront province of **George McKesson Brown**, heir to a pharmaceutical fortune. The house was designed to echo a 1700s chateau he had seen on trips to France.

When the Depression hit, however, Brown had to start selling the estate piecemeal. In 1938, he and his wife moved out of the mansion into the superintendent's residence. They sold furnishings they didn't take, then sold 33 acres with the mansion and boathouse to a Catholic teaching order — which added wings, glassed-in the interior courtyard, walled-over fireplaces and turned it into a boys school. The brothers renamed it **Coindre Hall** in honor of a priest who'd founded their order in France.

Suffolk County bought the property in 1971 and tried to put it to various uses. Finally, in 1989, it was transferred to the parks department, and tapped as a museum. After a faltering start, it's being restored via an ingenious coalition of government and private groups. A local rowing club's rental fee helps maintain the boathouse; a town recreation program pays the gym's utilities and cares for the north side grounds; catered events booked through Friends for Long Island's Heritage fund the upkeep of the south side gardens, and **Splashes of Hope**, a nonprofit organization that paints murals in hospital pediatric wards, has been given studio space in exchange for providing volunteer guides for mansion tours (offered for a donation most Saturdays and Thursdays at 11 a.m. and 1 p.m.).

Who says the Gold Coast can't rise again? ◆

Coindre Hall, once a home, was reincarnated as a school, then a museum.

GREENPORT

A Folksy World In One Square Mile

ANY TOWN that wants to attract tourists needs a catchy slogan, but the motto of one small North Fork village is actually true: **"Greenport is the real thing."**

Banners strung across the main road promote folksy summer events such as lobster festivals and ice cream socials, kids cannonball off small docks under the watchful eye of adults sitting on wide Victorian porches, and the whole town seems to turn out for the Friday night harborside band concerts held alongside the 1920s carousel. In the business district, seafood restaurants stand barely a clam's squirt away from the boats that supply the day's catch, and the eclectic non-chain shopping options range from an old-fashioned department store to the **"Last Winery Before France."** All this neatly within one square mile.

Greenport's timeline starts in the late 1600s, when the first European settlers called it **Winter Harbor**. Over the centuries, major industries ebbed and flowed, from whaling to oystering to shipbuilding — supplemented by a hefty splash of rum-running during Prohibition.

But its soul was always rooted in the sea, and many Greenporters still work at boat-related livelihoods — now as likely to involve tourists as cargoes of fish. Visitors can cruise the harbor on old-style sailing schooners or on an

Dockside dining at Claudio's Harborside Marina Clam Bar, a local favorite

electric-powered launch reminiscent of Venice's sleek, canopied vaporetti.

Greenport also was briefly the eastern terminus of the main line of the LIRR — in 1844 its **"Boston train"** became the fastest route between New York and Massachusetts (nine hours including a ferry link). By 1850, however, the daunting rivers and bays of Connecticut's shoreline were finally spanned by tracks, and Greenport settled back into its comfortable niche as a slow-lane seaport.

Pieces of its story are told via exhibits at the **East End Seaport Maritime Museum** and the **Railroad Museum of Long Island** — both near Greenport's LIRR station and Shelter Island Ferry dock, which passes for a mini-transportation hub. (For another perspective on the village's nautical past, you might visit the annual maritime festival, usually held in late September.)

Several of Greenport's oldest businesses have been operating since the late 19th century, and walking-tour brochures pinpoint these as well as numerous architectural landmarks notable for their previous lives. The 1894 **Greenport Auditorium**, for example, is a large Queen Anne structure that until 1938 was a center for highbrow entertainment intended to lift the cultural taste of sailors and shipbuilders.

Today, the hundreds of families who turn out for the Friday night sum-

Riding — and waiting to ride — the carousel in Greenport's Mitchell Park.

mertime concerts in Greenport's **Mitchell Park** come for more than free music by the all-volunteer band. They also like the accompaniments: the boats putt-putting across the harbor, the whirling carousel (its tunes respectfully muted during the concerts) and the toddlers gleefully dancing and somersaulting on the lawn. ◆

AT A GLANCE

The North Fork Tourist Information Center is in Southold, about a mile west of Greenport on Main Road (Route 25), 631-477-1383. **Hours**: Daily July and August; Friday to Monday, September and October; weekends April through June. Also check www.northfork.org and www.greenport.com. **Greenport's Stirling Historical Society** headquarters is at Main and Adams streets, 631-477-0099. **Hours**: 1 to 4 p.m. summer weekends. Partial wheelchair accessibility. Child appropriate.

Did You Know? Greenport was first called Winter Harbor, then Stirling (after revolutionary general William Alexander, known in Scotland as Lord Stirling), then Green Hill.

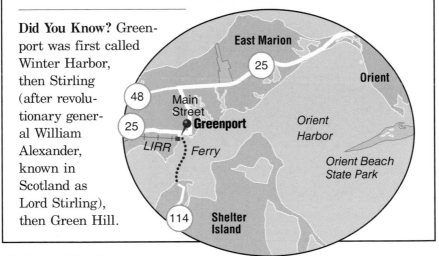

Take the Long Island Expressway to the last exit (Exit 73). Continue east on Route 58 (Old Country Road) for about four miles. Route 58 ends in Aquebogue and becomes Main Road or Route 25. Continue east on Route 25 for approximately 25 miles to Greenport.

◆

Ancient Woods, Modern Park

L IKE ANTIQUE BOOKENDS supporting your favorite escapist reading, two ancient forests protect the picnicking-swimming area of **Orient Beach State Park**.

On one side is a maritime red-cedar forest that is Long Island's only National Natural Landmark, a designation bestowed on such U.S. treasures as California's redwoods and Florida's Everglades. The forest grows on the

Newsday Photos / Tony Jerome

Long Beach Bar Lighthouse, off Orient Beach State Park, built in 1870

western half of the park's five-mile-long sandspit — which is punctuated by the now-offshore **Long Beach Bar Lighthouse** (whose squat shape has earned it the nickname The Bug Light). The landmark Long Beach area is off limits to boaters, but ambitious walkers can take guided hikes of the cedar forest or follow a self-guiding map (respecting seasonal beach closures designed to protect piping plovers and other nesting shore birds).

On the eastern side of the park's **Gardiners Bay** swimming-picnic area is a smaller maritime *oak* forest that is not globally rare, but still uncommon. Its trail has 13 numbered stops keyed to a brochure describing how plants and animals survive in this unusual environment — which has been submerged by hurricane floodwaters at least five times in the last century and also has withstood severe drought, bitter cold, insect infestation and timber-cutting. In one of nature's balancing acts, the ocean winds both severely stress the vegetation and supply it with needed nutrients carried in water droplets from the sea.

As well as some of the largest **blackjack oaks** on Long Island (which has America's northeasternmost population of these trees that have adapted

Summer visitors at Orient Beach State Park

to poor soil), you'll also see some **red cedars** (whose wax-coated, needle-like leaves protect them from the drying winds) and **prickly pear cactus** (whose water conservation devices are similar to the cedars'). Look carefully along the trail, too, for deer, rabbits and turtles — as well as various birds.

Had enough nature? There's swimming, fishing,

Along with swimming, hiking, picnicking and fishing, Orient Beach State Park offers a 2.5-mile bicycle path.

biking on a 2.5-mile path, boccie, volleyball, softball and horseshoes. Park regulars include a surprising number of city dwellers — who drive past all Long Island's other parks in favor of Orient's special low-key mix. ◆

Orient Beach State Park, eastern end of Route 25, 631-323-2440. Hours: Open daily year-round sunrise to sunset (swimming early June through Labor Day 10 a.m. to 6 p.m.). Fee: Bike rentals $4 an hour.

Newsday Photos / David L. Pokress

The Ann Currie-Bell Home, built about 1900 in the Victorian style

WHILE YOU'RE THERE

Oysterponds Historical Society, Village Lane, Orient, 631-323-2480. **Hours**: 2 to 5 p.m. Thursday, Saturday, Sunday mid-June to October. **Fee**: $3, 50 cents under age 10. Features a collection of 18th and 19th century buildings in Orient Village (once called Oysterponds). **Village House** was a 19th century boarding house and is furnished as such. **Hallock Building** holds maritime artifacts, paintings and agricultural tools. **Webb House**, an 18th century inn moved from Greenport, has period art and furniture.

Southold Historical Society, Main Road, Southold, 631-765-5500. **Hours**: 1 to 4 p.m. Wednesday, Saturday and Sunday early July to early September, except for the Prince Building, open 9:30 a.m. to 2:30 p.m.

Southold's Thomas Moore House

weekdays, year-round. **Fee**: Adults $2. Buildings include the **Ann Currie-Bell Home** with antique dolls and toys; the pre-1653 **Thomas Moore House** with furnishings from 1640 to 1840; and the 1821 **Old Bayview School**, which has been restored to its 1914 appearance. The 1874 **Prince Building** has records dating to the 17th century.

Cutchogue-New Suffolk Historical Council, Main Road, Cutchogue, 631-734-7122. **Hours**: 1 to 4 p.m. Saturday through Monday in July and August, and weekends in June and September. **Fee**: Donation. The best known in this grouping of historic buildings is **The Old House**, a national historic landmark that dates to 1649. This frame house was built in Southold and moved to Cutchogue in 1660. The

The Old House in Cutchogue

complex also includes the 1740 **Wickham House**, one of the oldest North Fork farmhouses, with a collection of quilts.

The **Southold Indian Museum**, 1080 Main Bayview Rd., Southold, 631-765-5577. **Hours**: 1:30 to 4:30 p.m. Sundays, plus Saturdays in July and August. **Fee**: $2; 50 cents under age 13. This museum, operated by the state Archeological Association's Long Island chapter, displays artifacts and relics of local Algonquians, as well as handiwork of many tribes of North and South America. Special programs and exhibits.

Custer Institute on Main Bayview Road in Southold, 631-765-2626. **Hours**: Open Saturdays year-round (call for hours and tours). **Fee**: Donation. This membership organization presents occasional lectures, classic films, concerts, art exhibits, as well as stargazing sessions at its astronomical observatory.

Horton Point Lighthouse, one of seven in Southold, has been guiding ships in Long Island Sound since 1857. **Hours**: The museum and tower are open 11:30 to 4 p.m. weekends Memorial Day through Columbus Day. **Fee**: Donation.

For a sampling of Greenport-Cutchogue restaurants, see Page 286.

Cultural Accents And Rebellious History

NATIVE SON Walt Whitman poetically described 19th century Huntington as "the place for him who wishes life in its flavor and its bloom."

It still is, Walt, rest assured.

The town's self-styled "Little Apple" billing is supported by a packed year-round cultural calendar. And when it really comes to celebrating itself, Huntington may have set its sights on outdoing its most famous resident on the occasion of its 350th birthday in 2003.

It's been an interesting few centuries — from the moment villagers on the brink of independence blew up an effigy of England's King George III to the time in the early 1900s when escape artist Harry Houdini broke out of the local jail (in one of his nationwide stunts to drum up business for his stage shows) to the strong bond with beloved 1970s troubadour Harry Chapin (who lived in Huntington, is buried there and is remembered via memorials and a charitable foundation) to the requisite link (however flimsier) to Long Island's favorite contemporary entertainer, **Billy Joel**.

The town was probably named for **Huntingdon**, England, birthplace of Oliver Cromwell — who was in power when it was founded in 1653. Huntington's own opposition to the crown escalated from the effigy-burning on the eve of the American Revolution to countless small victories (a local weaver's

The 1892 Soldiers and Sailors Memorial Building on Route 25A

The Huntington Arsenal, the only known colonial arsenal on Long Island

attic became area Patriots' secret munitions cache, a farmer's wife won respect when she popped her wedding ring into her mouth instead of handing it over to a menacing redcoat — prompting his officer to command him to "leave the plucky little lady alone"). There were losses, of course. Superpatriot **Nathan Hale** was captured sometime after landing his sloop in Huntington on a mission for Gen. George Washington. (History is murky on the details of his capture, or where he spent his final days before being hanged in Manhattan and supposedly uttering his famous sound bite, "I only regret that I have but one life to lose for my country.")

Fast forward to the 19th century, when the rural crossroads had become a notable address (the 1840 wedding of a prominent local minister's daughter to Cornelius Vanderbilt's eldest son inspired a society columnist to snipe that it was the nouveau riche industrialist who had mar-

Newsday Photo / Tony Jerome

The Long Island Philharmonic performs during a summer concert at Heckscher Park.

ried well). That century's war is remembered at the 1892 Soldier and Sailors Memorial Building, where visitors can see a tattered Civil War battle flag carried by a company of Huntington volunteers.

The historical society's **Conklin House** — begun about 1750 by the farmer with the plucky wife — is furnished to reflect a span of Colonial, Federal and Victorian times. Objects there include a chair and foot warmer perhaps used by George Washington at a local stop and a pump organ donated by piano man Billy Joel's mother (whoever might have played it, though, is unknown).

Speaking of popular culture, the **Huntington Arts Council** offers tons of exhibits, lectures, workshops, performances and other activities and events

AT A GLANCE

Huntington Township Chamber of Commerce, 631-423-6100 or www.huntingtonchamber.com. **Huntington Historical Society**, 209 Main St., 631-427-7045. Partially wheelchair accessible town-wide. Child appropriate.

Did You Know? The 1905 Huntington Sewing and Trade School (now historical society offices) was America's first vocational school.

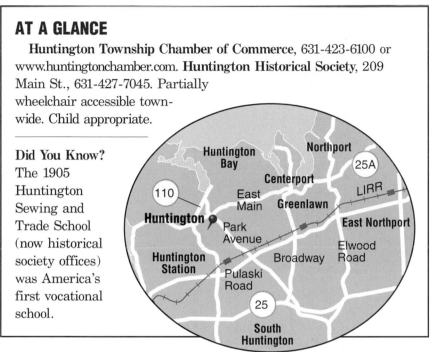

Take the Long Expressway to Exit 49N to
Route 110. Take Route 110 north approximately 8 miles to Route 25A in Huntington.
For Northport go right (east) on 25A about five miles.

for children and adults. Its Summer Arts Festival features three months of free music, dance and theater in the village's 18.5-acre downtown **Heckscher Park**.

The park's esteemed **Heckscher Museum of Art** boasts a permanent collection of more than 1,800 works spanning 500 years of western art from the European Renaissance painters to contemporary Americans — including many Long Islanders. As well as frequent special exhibits (each accompanied by a free down-to-earth guide designed for families), the museum also offers countless programs for kids and grown-ups including gallery talks, tours and workshops plus a Friday evening music series that yields in summer to the outdoor arts festival. (The park also is the site of a spring tulip festival, summer art show and fall festival of food, crafts and other fun stuff.)

Then there's the **Cinema Arts Centre, the Inter-Media Art Center** and other performing venues. Cosmopolitan shopping and dining? Natch. Residents must barely find time to put in a few hours a day at their jobs. ◆

Newsday Photo / Tony Jerome

A full moon still shines as dawn comes to Northport Harbor.

Trolley Tracks Into the Past

A S COASTAL LONG ISLAND shifted from farming to shipbuilding in the 1800s, the village formerly known as **Cow Harbor** was renamed Northport — no doubt a relief to status-conscious 19th century tourists.

The name was briefly changed once more, to **Greenleaf**, for the 1997 comedy "In and Out." Ostensibly, the stars were Tom Selleck and Kevin Kline, but locally the village cameos would have taken any Oscar. Northport has always had an astute appreciation for the arts.

It's also a gem of early Americana. Picturesque Main Street still has its trolley tracks, and you can hear old-fashioned band concerts in the park by the dock.

Hometown of "The Sopranos" star **Edie Falco** and singer-actress **Patti LuPone**, the village and adjacent Eatons Neck-Asharoken peninsula have long offered respite to a galaxy of stars. Enrico Caruso sang in Northport's Trinity Episcopal Church. In the 1930s, playwright **Eugene O'Neill** polished "Mourning Becomes Electra" in Eatons Neck. In the '40s, French writer and aviator Antoine de Saint-Exupery finished "The Little Prince"

Newsday Photo / Michael E. Ach

An ornate row of Northport buildings, erected between 1871 and 1895

in Asharoken, and Northporters rowed out to listen to composer **Sergei Rachmaninoff** at work across the harbor in Centerport. In the '50s, Jackie Gleason tooled around at the wheel of a motorboat trailing starlets on water skis, and a decade later, Marlene Dietrich could often be found at her daughter's house in Asharoken. Northport's mini-art colony included cartoonist-sculptor **Rube Goldberg** and artists Jules Olitski and Marcel Vertes. Even footloose writer **Jack Kerouac** put down temporary roots in the village, using profits from "On the Road" to buy three houses from 1958 to 1964 (they're now unmarked private residences, but you may run into his ghost in **Gunther's Bar**).

The historical society's "Walk Northport" brochure details numerous landmarks (don't miss the Victorian captain's row along Bayview Avenue), and its monthly walking tours add lots of inside stuff (its museum also offers programs and exhibits). As for shopping, you'll encounter as many villagers as tourists in the quaint-but-not-cutesy stores. Major events? Nothing is bigger than September's annual Cow Harbor Day Parade. ◆

Northport Chamber of Commerce, 631-754-3905 or www.northportny.com. Historical Society Museum, 215 Main St., 631-757-9859 or www.northport historical.org. Hours: Open Tuesday to Sunday 1 to 4:30 p.m.

WHILE YOU'RE THERE

Heckscher Museum of Art, 2 Prime Ave., 631-351-3250 or www.heckscher.org. **Hours**: Open Tuesday to Friday 10 a.m. to 5 p.m. (until 8:30 p.m. the first Friday of the month), weekends 1 to 5 p.m. **Fee**: Voluntary suggested donation $3 adults, $1 younger than age 12. The neoclassical building was erected in 1920 by August Heckscher to house his private art collection.

The Arsenal, 425 Park Ave., 631-351-3244. **Hours**: Open most Sundays 1 to 4 p.m. **Fee**: Free. This site was constructed circa 1740 as a farm building, then became home to weaver Job Sammis, who volunteered his attic as a weapons cache for Revolutionary militiamen who drilled on the nearby **Village Green.** Highlights include displays on the life of a working family of the period, plus a collection of muskets.

Huntington Historical Society Museums (for information on the following three sites, call 631-427- 7045): **Kissam House**, 434 Park Ave. **Hours**: Open Sundays 1 to 4 p.m. May through October. **Fee**: $2.50 adults, $2 seniors, $1 under age 12, $5 family rate. Built in 1795 for Dr. Daniel Whitehead Kissam, one of the town's early physicians (whose young relative later made headlines when she married the Vanderbilt), it displays period furnishings (family pieces include a circa 1810 Seth Thomas tall-case clock and the doctor's medical diploma). An upstairs room is devoted to the family's history. The barnyard is the site of an annual

On the Village Green, the Huntington Militia recalls the start of the American Revolution. At right, a topsy-turvy doll on display at the Conklin House on High Street.

Newsday Photo / David L. Pokress

spring Sheep to Shawl Festival and fall apple festival; the barn houses a consignment shop (and soon re-created leather and woodworking shops, as well). **Conklin House**, 2 High St. at New York Avenue (Route 110). **Hours**: Open year-round Tuesday to Friday and Sunday 1 to 4 p.m. **Fee**: House tour $2.50 adults, $2 seniors, $1 under age 12, $5 family; exhibit room free. Typical Long Island farmhouse started about 1750 and revised over the years by the family of David Conklin, a Patriot who fled to Connecticut at the start of the Revolution but returned after his wife was threatened by a British soldier when she refused to hand over her wedding ring (Conklin was alternately imprisoned, then forced to help the redcoats build two nearby forts and prepare meals for soldiers billeted in the village; he died a few years after the British left Huntington). Furnishings from colonial to Victorian portray its occupancy by numerous generations. **Soldiers and Sailors Memorial Building**, 228 Main St. **Hours**: Open 1 to 4 p.m. Sundays May through October. This site was dedicated in 1892 as a library and Civil War memorial; now it displays a restored Civil War battle flag carried by the 127th Regiment New York State Volunteers' Huntington company.

Walt Whitman Birthplace State Historic Site, 246 Old Walt Whitman Rd., Huntington Station, 631-427-5240 (for more details, see Page 259).

For a sampling of nearby restaurants, see Huntington on Page 287 and Northport-Greenlawn on Page 289.

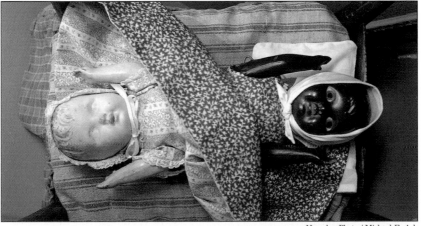

Newsday Photo / Michael E. Ach

JOHN P. HUMES
JAPANESE STROLL GARDEN

Taking The Path To Enlightenment

A S PROFOUND as a wisp of haiku verse, the tiniest Japanese stroll garden is designed to subtly lead visitors on a mini-journey toward enlightenment.

Every pebbled path and stepping-stone, every weeping willow and grove of swaying bamboo, plays a part in the overall plan to immerse you so completely in nature that you become one with it — eventually discovering the not-so-fast track to inner peace.

And you thought this was going to be just another everyday walk in the woods.

"You *are* the garden while you're here — you're as natural a part of this garden as the bird that just landed in the tree above us, or the water, or the plants," explains curator Stephen Morrell on tours of the **John P. Humes Japanese Stroll Garden** in Mill Neck.

This unusual local refuge from the world was inspired by a 1960 visit to Japan's ancient capital of Kyoto by lawyer (and later ambassador to Austria) **John P. Humes.** As soon as he got home he hired a Japanese landscape gardener, and within four years a two-acre corner of his Long Island estate was transformed into an ink-brush landscape complete with an im-

The carefully placed stepping-stones at the Humes Japanese Stroll Garden

ported teahouse. Humes' Japanese garden fell into disrepair while he was abroad on his diplomatic assignment but when he returned, in ill health himself, he engaged Morrell to rehabilitate it. In 1986, a year after Humes died, it was opened to the public.

You can learn a lot from the brochure, but a walk led by Morrell is the optimum introduction to this tranquil oasis, which now covers close to four acres — blending gardens and native woodlands with religious and philosophical principles of both **Shinto** and **Zen Buddhism** (embraced simultaneously by many Japanese).

Curved pathways suggest the twists and turns of life.

"The intent is a walking meditation," Morrell says, describing how the stepping-stones set in the path control the rate at which visitors move through the garden: Small stones slow you down, larger ones invite you to plant both feet and take in a particular vista.

Your journey begins at the first of three garden gateways, near a posted inscription: "May peace prevail on earth." A dirt trail winds across a narrow log bridge, then zigzags up a conceptual **"seaside mountain"** (actually a gentle suburban hill). This represents the struggle toward enlightenment: During the climb, you can't see what's around the next bend but are meant to wander and wonder.

When you reach the second gate, atop the mountain, you suddenly see more clearly — which is what meditation is meant to do for the mind. The natural woodland also morphs into more cultivated surroundings, adorned with stone lanterns and other symbolic objects. And the descending paths are pebbled — representing streams cascading downhill over symbolic waterfalls (there's a real waterfall, too) to the "ocean." Here the sea is a pond harboring goldfish-like **Japanese**

AT A GLANCE

John P. Humes Japanese Stroll Garden, corner of Oyster Bay Road and Dogwood Lane, Mill Neck, 516-676-4486. **Hours:** 11:30 a.m. to 4:30 p.m. weekends through late October. **Fee:** Entrance fee $5; younger than age 12 free. Guided tours with Japanese tea ceremony demonstration 10 a.m. some Saturdays ($10, reservation required). Other programs include shakuhachi flute concerts and meditation sessions, as well as ikebana (Japanese flower arranging) workshops. Partially child appropriate.

Did You Know? The Zen arts, including gardening, use techniques such as asymmetry to compel individuals to change in response to environmental changes.

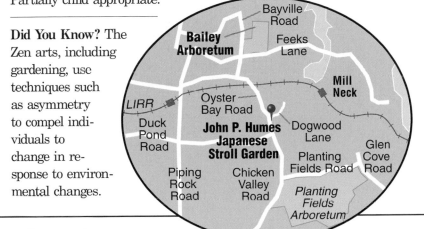

From the Long Island Expressway Exit 41N, take Route 107 (Cedar Swamp Road) north about three miles to Wolver Hollow Road on right. Turn right onto Wolver Hollow, drive about three miles and turn right (north) onto Chicken Valley Road. After passing Planting Fields Arboretum on the right, Chicken Valley becomes Oyster Bay Road. Take Oyster Bay Road a short distance to Dogwood Lane on the right. The John P. Humes Japanese Stroll Garden is at the intersection.

A footbridge on the Humes garden's symbolic path to enlightenment

The steady sound of falling water invites contemplation.

koi, turtles and catfish (including a burly black one called Hoover because, in a very un-Zen spirit, at feeding time he vacuums up every morsel meant to be shared).

Central to the garden journey is the teahouse, set amid evergreens and nonshowy plants that won't distract from the tea ceremony — which embodies a spiritual exchange between the host, an honored guest and the environment. The third gateway, outside the tea garden, symbolizes an enlightened return to earth.

The painstaking formality of the tea ceremony — which Morrell demonstrates with a tour-group volunteer — soon has some Type-A tourists furtively glancing at their watches. But Morrell later explains that the ritual is meant to take something ordinary and make every detail matter — a lesson in giving yourself completely to every moment in life. "In the Zen arts," he says, "the only goal is to be where you are."

Wherever that may be after your garden visit, and whatever worldly cares await outside, chances are you'll long remember the harmony and serenity within. ◆

Serenity Amid the Woodlands

T OWERING TREES, including Long Island's largest dawn redwood and black walnut, seem poised to brush away any clouds that happen to drift over Lattingtown's **Bailey Arboretum**. Another skysweeper — an elegant black locust, its trunk swathed in a summer-white cloak of climbing hydrangea blossoms — stands watch there, too, just off the paved walk behind the main house.

These rare specimens are among the more than 600 varieties of trees and shrubs from around the world that dot the 42-acre former summer estate of New Yorkers **Frank and Louise Bailey** — who wryly referred to it as **"Munnysunk."** But their passion for collecting exotic plants was clearly money well spent — in view of the pleasure they gained from their pastime and in the continuing enjoyment of visitors to the property since 1968, when it was acquired by Nassau County.

In addition to trees such as dwarf Nikko firs from Japan, blue atlas cedars from North Africa, Chinese dawn redwoods and Korean pines, there's also a **three-season flower display**. Annuals, perennials and flower-

Newsday Photos / Ken Spencer

Munnysunk, the estate house at Bailey Arboretum in Lattingtown

ing shrubs — highlighted by irises, roses, rhododendrons, azaleas and chrysanthemums — make an impressive splash against the manicured lawns (beware of strolling in new shoes, however, as the profusion of Canada geese can create a minefield, of sorts).

The estate house, built in the mid-1800s, isn't regularly open to the public, but is available for use by garden clubs and conservation-related organizations. Beyond the manicured yard with its international transplants, the grounds remain in a natural state — with trails winding through woods distinguished by century-old tulip, oak and maple trees that are Long Island natives (the map-brochure also pinpoints the island's largest tricolor beech). In spring, the forest floor is as colorful as the backyard — carpeted with violets, trout lilies, trilliums, May apples and jack-in-the-pulpits.

To complete the something-for-everyone design, a raised sensory garden featuring plants that can be experienced not just by sight but also by smell, taste and touch is located just off the parking lot. ◆

Bailey Arboretum, Bayville Road and Feeks Lane, Lattingtown, 516-571-8020. Hours: Tuesday to Sunday mid-April through mid-November (hours vary seasonally). Fee: Nassau County residents $1, nonresidents $3.

A 2001 sign welcomes visitors to Muttontown Preserve. Above, a quiet scene at Church Nature Sanctuary / Shu Swamp.

WHILE YOU'RE THERE

Muttontown Preserve, Muttontown Lane (south off Route 25A one-tenth of a mile west of Route 106), 516-571-8500. **Hours**: Open daily 9:30 a.m. to 4:30 p.m. **Fee**: Free. This 550-acre Nassau County park of fields and woods includes some interpretive placards on local wildflowers, trees and animals along a "self-guiding trail" outlined in the map-brochures available at the preserve's Nature Center. Main trails are marked with colored-number posts, but you're guaranteed to briefly get lost at least once in the maze of side paths — many no narrower than the main ones. Since this is horse country, though, you can ask directions of horseback riders you're likely to meet (there's an equestrian entrance with horse-van parking at the preserve's southern end, off Route 106). Cross-country skiing in the winter.

Charles T. Church Nature Sanctuary / Shu Swamp, Frost Mill Road, Mill Neck, free. This site offers 2.5 miles of trails, with boardwalks across muddier areas.

For a sampling of nearby restaurants, see Glen Cove on Page 284, Locust Valley on Page 288 and Oyster Bay on Page 290.

JONES BEACH

Eight Ocean Beaches Called Jones

'TWO HOT DOGS, please — hold the sand."

The young wag in the Jones Beach grill line had to be joking. Everyone knows the sand isn't added at the snack bars. It's sprinkled on by the gentle breeze and the hundreds of sizzling feet that dance past your blanket while you're savoring one of life's greatest pleasures: an oceanside junk-food lunch.

Lunchtime entertainment? Just look around you. Sunbathers preen on a rainbow of towels, boogie boarders coast atop the breakers rolling in from Africa, toddlers intently scoop holes at the water's edge, and watchful elders knee-deep in the frothy backwash reminisce about days when they, too, tried to dig to China and built castles in the sand. Life is indeed a beach — especially a beach named Jones.

Jones Beach is actually eight ocean beaches strung together along a 6.5-mile sandspit, plus a half-mile still-water beachfront on **Zach's Bay**. They're just part of a 2,413-acre state park that also offers — sometimes limited by area, season or permit — two pools, surfing, surf and bay fishing (day and night), stargazing, a band shell with outdoor dancing, a 14,000-plus-seat outdoor theater whose summer concert schedule includes top stars, a 1.9-mile boardwalk dotted with photos of early beachgoers in shin-to-chin bathing costumes or business suits and straw hats (all Long

Visitors arrive at the Central Mall to begin a day at Jones Beach.

Island state parks are highlighted in the **Castles in the Sand** historical exhibit at the East Bathhouse), a two-mile health walk, a boat basin, picnic areas, basketball and shuffleboard courts, softball fields, paddle tennis, volleyball, 18-hole pitch-putt golf, miniature golf, playgrounds, a nature center, gift shops and a glass-walled full-service restaurant.

Skateboarding, roller-skating and in-line skating are prohibited in the park year-round, and bicycling is banned from the boardwalk April 1 to Oct. 1, but they're allowed on the 4.5-mile recreation path running along Wantagh State Parkway from Cedar Creek County Park on Merrick Road to Parking Field 5 at the Jones Beach Theater (where there are bike racks).

The state park's two lofty landmarks nearby are more than decorative. The 231-foot **Jones Beach Tower**, modeled after the campanile of St. Mark's Cathedral in Venice, stores 316,000 gallons of water from wells that supply the park. And the bright signal flags that fly from the nearby Central Mall's 90-foot ship's mast flagpole bear the coded messages **"Jones Beach State Park"** and "Keep Your Park Neat," followed by the current year.

Newsday Photo / Michael E. Ach

Swimmers enjoy the Jones Beach surf.

Jones Beach opened Aug. 4, 1929, during a voice-choking, car-stalling sandstorm that almost canceled the ceremony marking the completion of a troubled project. Voters defeated the first referendum proposed by the Long Island State Park Commission's irrepressible president, **Robert Moses** (rumors that the proposed public beach would have no local benefit reportedly were started by wealthy estate owners who feared a Coney Island East). Then labor strikes stopped work awhile. And money became such an issue, Moses had to persuade his mother to advance $20,000.

Despite critics, Jones Beach has always been a crowd pleaser. It drew 1.5 million visitors its first year and now logs 6 million to 8 million annually — most during summer. The all-time one-day record tally was 276,000, on July 4, 1998. (Beachgoers, incidentally, annually scarf up 1.5 million hot dogs, 650,000 hamburgers and 1.2 million servings of ice cream.)

Today, there are official radio-free zones as well as unofficial favored gathering spots. Senior citizens and families like **Field 6**, for example, because it

AT A GLANCE

Jones Beach State Park, Ocean Drive (southern end of Meadow-brook and Wantagh parkways), Wantagh, 516-785-1600, www.nys parks.com.

Hours: Sunrise to midnight from third weekend in June until Labor Day, rest of the year sunrise to sunset; West Bathhouse open third weekend in June until Labor Day 10 a.m. to 6 p.m. weekdays, 9 a.m. to 8 p.m. weekends and holidays (until 7 p.m. mid-August to Labor Day); East Bathhouse open 9 a.m. to 8 p.m. weekends and holidays only until mid-August.

Fee: $7 per car daily late May to mid-September, $5 weekends and holidays early April to mid-May and late September through November; extra charge for pools and many activities. Wheelchair accessibile. Child appropriate.

Did You Know? The park was named for Maj. Thomas Jones, a soldier and seaman who settled nearby in 1692 and set up a whaling station.

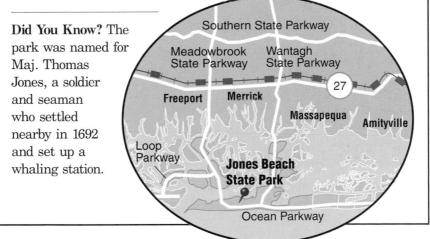

From the Northern State Parkway take
either the Meadowbrook Parkway south at Exit 31A or the Wantagh State Parkway south at Exit 33.
Take either parkway south to end at Jones Beach State Park. From the Southern State Parkway take
the Meadowbrook south at Exit 22, or the Wantagh south at Exit 27 to Jones Beach State Park.

has the shortest walk from car to water (young families also like the Zach's Bay beach, where there's a day camp weekdays). Nature lovers prefer West End **Field 2**, which has the longest walk across a sparser-peopled beach and thus more solitude. Teens and post-teens, of course, immediately head for the busy Central Mall. ◆

It's Only Natural

T HE FIRST CLUE is the sign: "Please No Climbing on Bones!" This clearly isn't your average sandbox.

It's the Mystery Bone Discovery Cove at Jones Beach's Theodore Roosevelt Nature Center, an educational playground that's home to some mighty big vertebrae and other skeletal parts.

A dozen 6-year-olds diligently digging amid these half-buried treasures recently paused to ponder an instructor's question. Whose bones are these? Some thought a dinosaur's. But the majority answered correctly: The "bones" (actually fiberglass models) are a whale's. And inside the nature center building are some authentic artifacts — not just whale bones but also books and exhibits featuring shore birds, animals and butterflies. A frequently changed assortment of live-but-harmless sea creatures inhabit a Touch Tank ("When we find something really cool like a skate or a turtle that could bite, then it becomes a Do Not Touch Tank," assured director Annie McIntyre).

Newsday Photo / Bill Davis

Local animal life on display at the Theodore Roosevelt Nature Center at Jones Beach.

Other attractions outside include a butterfly garden filled with plants irresistible to the familiar orange and black monarchs (two of 1,000 tagged here were recorded at a winter migratory stopover in Mexico), a human sundial (your body throws the shadow that indicates the time), the outline of a blue whale (done with blue paint, of course, and measuring 53 adult paces long) and the hulk of a shipwrecked barge (which sank off Jones Beach in 1895 and washed ashore 98 years later). A boardwalk across the environmentally fragile dunes should be complete in 2002. It will offer observation points closer to the off-limits nesting areas of birds such as piping plovers and least terns, though there's no guarantee they'll always nest with optimum viewing in mind.

WHILE YOU'RE THERE

Newsday Photo / Daniel Goodrich

A classic parlor car at the Historical and Preservation Society Museum

When you've had enough sun and surf, if it's a Sunday afternoon you might stop by the **Wantagh Historical and Preservation Society Museum**, 1700 Wantagh Ave. (between Jerusalem Avenue and Sunrise Highway), Wantagh, 516-826-8767. **Hours**: Sunday 2 to 4 p.m. April through October. Closed November through March. **Fee**: None. The restored 1885 former Wantagh railroad station includes history displays and, outside, a 1912 LIRR parlor car and other exhibits.

Massapequa Historic Complex, Merrick Road at Cedar Shore Drive, 516-799-2023, www.massapequahistory.com, includes a book-filled 1896 library, a cozy circa-1870 servants cottage and the Gothic 1844 Old Grace Church (in its burial ground is the grave of Maj. Thomas Jones, the town's first non-Indian settler, for whom Jones Beach was named). **Hours**: Library year-round 10 a.m. to 1 p.m. Wednesdays and Saturdays. Church and cottage May through September, 2 to 4 p.m. Sundays (other times by appointment). **Fee**: Free.

Tackapausha Museum, Washington Avenue, Seaford, 516-571-7443. Tackapausha Preserve. **Hours**: 10 a.m. to 4 p.m. Tuesday through Saturday, 1 to 4 p.m. Sunday, year-round. **Fee**: $2; ages 5 to 12 $1. Seasonal exhibits of Long Island wildlife and plants, with a nocturnal animal exhibit, are used to interpret the principles of life sciences at this 80-acre preserve. There are also five miles of nature trails.

For a sampling of nearby restaurants, see Freeport on Page 283 and Long Beach area on Page 288.

The center offers indoor and outdoor programs year-round — some for school and youth groups, others geared to children, families or adults. ◆

Theodore Roosevelt Nature Center, Jones Beach State Park, 516-679-7254. Hours: 10 a.m. to 4 p.m. Wednesday through Sunday Memorial Day to Labor Day, weekends year-round. Fee: Free; programs $3 per person.

LONG BEACH

'See You On The Boardwalk'

ALL THOSE STALKS of steamed asparagus who flip-flop off the Manhattan express trains on sultry summer weekends come to **Long Beach** for one thing: its world-class slice of ocean five blocks from the LIRR.

But when escaping a heat wave isn't the sole objective of a visit, day-trippers can savor the illusive qualities that have prompted generations of natives to proclaim: "There's Long Beach sand in my shoes."

One is longtime City Council leader Joel Crystal, whose favorite spot for 50 years has been "**Neptune Beach** between 4:30 and 7 p.m., when everybody's gone home, the sun's setting, and there are just the birds, the surf, the peace and quiet." Even his wiseacre brother Billy, who left for the bright lights of Hollywood, invariably becomes sentimental when interviews turn to his Long Beach childhood. Other celebrities have been habitues over the years, and, for one brief moment in its boom-and-bust past, the community boasted the world's largest seaside hotel (soon outdone by another elsewhere). But Long Beach takes it all in stride.

Although it's one of Long Island's only two cities (the other is Glen Cove), Long Beach has the casual aura of a small Florida or California beach town: stucco homes painted white or tropical pastels, small gardens displaying the sort of spiky plants common in the South or West, breezes

Long Beach's 2.2-mile boardwalk is a magnet for cyclists, walkers, joggers and others in a resort that once called itself "America's Healthiest City."

fragrant with suntan lotion. There's also a mix of cooking aromas befitting a cultural bouillabaisse — which here supports a variety of noteworthy fusion restaurants. The sandy lanes of bygone days may have morphed into wide boulevards, but quaint areas remain, such as **The Walks**, a group of homes accessible only by footpaths.

Some of the ocean-view apartment buildings still have a number of elderly residents — many drawn by the motto coined by a doctor in the 1920s: "America's Healthiest City" (the more modest **"City by the Sea"** now adorns the official seal). But with a renaissance seemingly always underway, this is hardly a doddering town.

Witness the vigorous year-round activity along the 2.2-mile boardwalk. Strollers, power-walkers and riders of bicycles built for one, two, three or four regularly ply the boards. You also can have someone pedal you around, in Long Beach's version of Atlantic City's rolling carts. Lots of benches beckon those who want to pause to watch the action on both the boardwalk and the beach — a vantage surpassed only by the lifeguards' perches atop mini-mountains of sand.

The boardwalk has only one small snack bar complex — about dead center, between **National** and **Edwards boulevards** — which everyone seems to consider perfectly adequate. After all, there are plenty of eateries just a few blocks' walk. This keeps the city's ambience more residential than resort.

In addition to the town center along Park Avenue, there's a smaller commercial district in the West End, which was an early bungalow colony laid out along streets named for states. This neighborhood has most of the city's few nightspots — including **Shines**, the remaining speakeasy of an original 28 (more bars and clubs are across Reynolds Channel in Island Park). On the channel side of Long Beach, there's a fishing pier (at Magnolia Boulevard and West Bay Drive) and Hempstead Town marinas to the east (one on either side of the Meadowbrook Parkway access road).

Long Beach has a full slate of festivals, concerts, volleyball and sand-soccer tournaments, swimming and running races. Such activities wind down offseason, when the focus turns to arts, ethnic and historical events. Major landmarks are noted on a see-it-yourself-anytime map that supplements the annual guided heritage tours.

Post-summer events are generally held away from the windswept waterfront, but the prevalent year-round parting words remain: "I'll see you on the boardwalk." ◆

Newsday Photo / Dick Kraus

On a July day, a surfer surveys the Atlantic waves near Edwards Avenue.

AT A GLANCE

Fee: From Memorial Day Weekend to Labor Day, everyone older than 12 must pay ($6 daily without a season pass) to use the beach. Swimming only when lifeguards are on duty (8 a.m. to 6 p.m.). Surfing areas are designated in the East and West Ends. Street parking is free, with a time limit in some business areas. There's a seasonal information center on the boardwalk at Riverside Boulevard. For year-round information, call City Hall: 516-431-1000. **Long Beach Historical Society**, 226 W. Penn St., 516-432-1192. **Hours**: 9 a.m. to noon Monday to Friday. Partially wheelchair accessible. Child appropriate.

Newsday Photo / Michael E. Ach

A 9-month-old boy is introduced to the Atlantic Ocean at Long Beach.

Did You Know? The legend that elephants built Long Beach's boardwalk may stem from the fact they helped demolish cottages that were in the way.

Take the Meadowbrook State Parkway south to Exit M10. Take the Loop Parkway to Lido Boulevard in Lido Beach. Go west on Lido Boulevard about two miles to Long Beach.

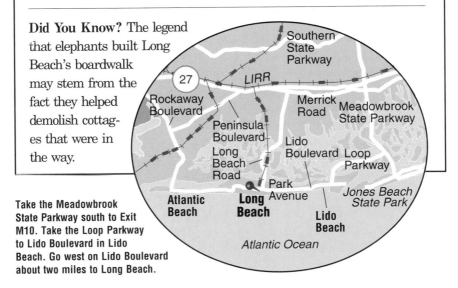

Newsday Photo / Dick Kraus

The Lido Towers, formerly the Lido Beach Hotel

Echoes of a Bygone Era

L ONG BEACH was valued by tourists as far back as the 1600s, when Hempstead settlers bartered with the sachem of several South Shore tribes to share the "long white sandy beach."

More than 200 years passed before the first house was built there (and promptly washed away). But development persisted, and in 1880 the narrow barrier isle gained both a rail link and what was for several years the world's largest seaside resort: the 101-room **Long Beach Hotel**. When that burned in 1907, developer William Reynolds built a new fireproof hotel on the same spot. The next year, a boardwalk (first made of concrete) was built. So were The Estates of Long Beach — Reynolds' planned community of white stucco homes with red tile roofs along red brick streets, which he envisioned as a second Atlantic City (pre-casinos).

One of the original estate houses is now home to the **Long Beach Historical Society**, which displays a small collection of photos and artifacts from bygone days, sponsors an annual guided heritage house tour and other events, and publishes a walking-tour map. Sites include two Depression-era National Landmarks: the U.S. Post Office (listed for its Jon Corbino mural "The Pleasures of the Bathing Beach") and the Moorish-style **Granada Towers**, Nassau County's first luxury apartment house (which opened the day of the 1929 stock market crash). Placards also mark locations around town

WHILE YOU'RE THERE

Oceanside Marine Nature Study Area, 500 Slice Dr., Oceanside. Operated by the Town of Hempstead, 516-766-1580. **Hours**: 9 a.m. to 5 p.m. Tuesday to Saturday year-round, except holidays. **Fee**: Free; displays, trails on 52 acres, mainly salt marsh; bird-watching.

Nassau Beach Park, Lido Beach, 516-571-7700. **Hours**: Daily in summer, weekends only late May to mid-June. **Fee**: Parking $6 Nassau residents, $17 nonresidents. Facilities include two pools and tennis courts (fee for both), ball fields, playgrounds, campsites, picnic area (reserve) and surfcasting (permit).

Bay Park, Fifth Street, East Rockaway, 516-571-7245 or 516-571-7821 (offseason). Call for hours and fees. This 96-acre Nassau County park features a nine-hole golf course, ball fields, cricket fields, fishing dock, picnic area, playground, outdoor roller rink and a dog run.

Grant Park, Sheridan Avenue, Hewlett, 516-571-7821. Call for hours and fees. Facilities at this 35-acre Nassau County park include four tennis courts, three softball fields, roller-skating and ice-skating with rentals (fee), picnic areas and a fishing lake.

For a sampling of Long Beach area restaurants, see Page 288.

where famous landmarks once stood — such as the Long Beach Hotel and its similarly Queen Anne-style "cottages" (rented to prominent summer visitors). Two 1881 London plane trees are labeled the oldest living veterans of 19th century life.

Only one of the grand hotels from these early glory days survives. The **Lido Towers**, now a luxury condo, lacks some distinctive features of its original incarnation as the Lido Beach Hotel. Built in 1928 during the second phase of Reynolds' development of Long Beach — this time intended to re-create Venice, right down to the canals still gracing the East End north of Park Avenue — the Lido Beach boasted a sky roof that opened for dancing under the stars. (An annual summer boardwalk parade of antique and classic cars re-creates some of the atmosphere of that era.)

Reynolds' personal fortunes were as bumpy as the city's. He was elected its first mayor, then jailed in its first corruption scandal. But a judge ultimately dismissed all indictments against him — a verdict today's Long Beach boosters undoubtedly would cheer. ◆

LONG ISLAND MARITIME MUSEUM

Setting Sail Into The Seafaring Past

THE ORIGINAL sail-powered Fire Island ferry.

A boat crafted by a freed slave, who named it Abe Lincoln to honor the president who'd issued the Emancipation Proclamation.

A child-sized square-rigger built to ensure that two young shipping line heirs would be experienced seamen when they took the company's helm.

These are just a few of several dozen boats — all crafted locally and accompanied by compelling tales — that fill the small-craft exhibition building and line the dock at the **Long Island Maritime Museum** in West Sayville. In its boat shop — a historic building floated by barge to its present site — are more. Some are being restored by skilled veterans, others are being shaped under their supervision by a new generation of boatbuilders that may one day include seafaring legends equal to the revered **Gil Smith** (the museum collection includes a dozen of his designs).

Boats, you see, aren't a static focus at this absorbing indoor-outdoor museum. In its mission to preserve local maritime history, it also spotlights the lifestyle and crafts of the people who for generations have earned a living from local waters.

At the turn of the last century, Great South Bay was considered the

Scouts learn about boat construction at the West Sayville museum.

most bountiful oyster ground north of Chesapeake Bay — known world-wide for its succulent, plump bluepoints. The **William Rudolph Oyster House**, which played a key role in the local industry from 1908 when it was built, to the 1940s when the Rudolph business dissolved, is the only such 19th century relic left on Long Island (moved from its original location on Shore Road to the museum grounds). Visitors feel that they've stopped by in the midst of the work day — the long "opening bench" piled high with oyster shells and scattered with tools of the trade.

As early as the 1870s, unrelenting harvesting brought signs of exhaustion to the natural oyster beds and companies began to plant and cultivate oysters on private grounds. Individual baymen, however, could usually only afford to work on the natural public oyster beds. The multimillion-dollar oyster industry peaked in the 1890s and gradually declined until it finally ended in the 1950s, a victim of changes in the bay's chemistry caused both by pollution and storms that opened new inlets. Proudly docked in the museum's boat basin is the restored gaff-rig sloop **Priscilla**, the only sailing work boat surviving from the Great South Bay oyster fleet. "Pris" has participated in various regattas for classic vessels, including a parade of tall ships in New York Harbor.

Across the lawn is the modest cottage built by the **Beebe family**, among the first Dutch immigrants who came to the area in the late 1800s, helped develop the oyster industry and remain in the neighborhood. The Beebes lived in the house until 1982, before it was moved from its original site just up the street to join the complex of museum buildings. Today's visiting school groups get a tour by one of the pioneering Beebes' great-grandchildren.

The main museum building was the carriage house of an estate called **Meadow Edge** (former home of Florence Bourne Hard, whose father was the president of Singer Sewing Machine Co.). It's filled with maritime memorabilia, plus a permanent exhibit honoring the **U.S. Lifesaving Service** — which also charts the several hundred shipwrecks and distress calls that occurred off Long Island's South Shore from 1657 to 1996. Yearlong changing exhibits on tap are model ships in 2002, maritime instruments and devices in 2003 and maritime folk art in 2004. For true old salts, a reading room is always stocked to the gunwales.

As for special events, the hands-down favorite is the annual end-of-summer **Long Island Seafood Festival**. Then everyone with a taste for the sea heads to the museum for succulent home-cooked fare ranging from boiled lobsters to, of course, fried oysters. ◆

Newsday Photo / Katherine C. Martin

The Charlotte, a restored century-old tug that once worked in Northport Harbor, is among the boats on exhibit at the maritime museum.

AT A GLANCE

Long Island Maritime Museum, 86 West Ave., West Sayville, 631-854-4974. **Hours**: Monday to Saturday 10 a.m. to 4 p.m., Sunday noon to 4 p.m. **Fee**: Free ($4 suggested donation). Partially wheelchair accessible. Child appropriate.

Did You Know?
Among the many laws passed to preserve the oyster supply was a ban on harvesting them during the annual spawning season (months without an "r").

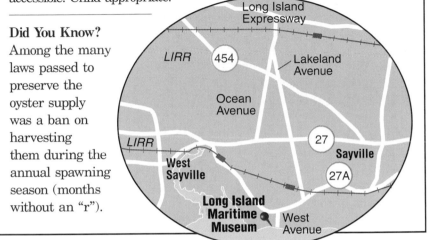

Take the Long Island Expressway to Exit 59S to Ocean Avenue. Go south on Ocean Avenue about two miles, then bear left onto Lakeland Avenue. Continue south on Lakeland for about seven miles to Montauk Highway (Route 27A) in Sayville. Make a right (going west) onto Montauk Highway and drive about two miles to West Avenue. Make a left onto West and proceed to museum.

The South Shore Roosevelts

L ONG ISLAND'S OTHER Roosevelts established their country compound on the South Shore, about 30 miles from the Oyster Bay stamping grounds of President Theodore Roosevelt's father and two other uncles.

Whatever prompted the geographical separation (you can ask about theories on your visit), T.R. remained tight with the Sayville contingent. One infamous day in 1903, the president, two of his sons and assorted nephews headed off on horseback at sunset to pay a visit. They rode all night, arriving for breakfast (the Secret Service, apparently less enthusiastic about

Newsday Photo / Katherine C. Martin

Meadow Croft in Sayville, home of John Ellis Roosevelt, a cousin of Theodore Roosevelt

roughing it, got there later by train).

The only home left on the original compound is that of **John Ellis Roosevelt** (who served as legal adviser to his cousin, the president). Called **Meadow Croft**, it now sits on 65 acres amid Suffolk County's **Sans Souci Lakes County Nature Preserve** on the Bayport line.

John bought the unpretentious 1860s farmhouse in 1890 and commissioned local architect **Isaac Henry Green** to design additions and Victorian porches typical of the period's "transitional style." The property's "auto house" and in-ground swimming pool (both now in disrepair) were among the state's first. However, the restored carriage house studio where John pursued his painting hobby is today used by **Splashes of Hope** — an artists' group that paints cheery murals in hospitals.

The 23-room house — still very much a work in progress — showcases a few family heirlooms and a lot of Old World craftsmanship (the carpenter purposely installed one stair-rail spindle upside-down, saying that only God can make something perfect). There are also displays of period clothing and Bayport memorabilia.

Walk or drive past the house on the dirt road and you come to tiny **Loughlin Vineyards** — whose owner, Barney Loughlin, was born and raised on the estate while his parents were the Roosevelts' nanny and caretaker. The annual harvest is done in one day with the help of friends and neighbors (average yield: 10 tons of handpicked fruit). Weekends in season, there's a mini-tasting bar at the edge of the 5-acre field. The donkey on the label? Pinky, a longtime family pet. ◆

Meadow Croft, Middle Road, Sayville, 631-472-4625. Hours: Sundays mid-May to mid-October; tours at 1 and 3 p.m. Fee: Free (donations accepted). Loughlin Vineyards, 800-803-0447. Hours: Weekends in season, noon to 6 p.m.

WHILE YOU'RE THERE

Bayport Aerodrome Society's Living Aviation Museum, Vitamin Drive, off Church Street, 631-277-2229 (call in advance). **Hours**: Noon to 4 p.m. Saturday and Sunday March through December, weather permitting. Most Sunday afternoons, visitors will find aficionados of vintage aircraft flying and restoring their planes (some rated for aerobatics) on and around the grass airstrip at Islip's old Edwards Airport. Among the dozens of intriguing curiosities: two British Tiger Moths whose propellers rotate in the opposite direction to a U.S. aircraft's and a Stinson 105 adapted with "ballistic parachutes" that could be deployed after jettisoning the wings if the engine quit.

Newsday Photo / Michael E. Ach

Antique-aircraft enthusiasts at the Bayport Aerodrome

Bellport-Brookhaven Historical Society, 31 Bellport Lane, Bellport, 631-286-0888. **Hours**: Museum 1 to 4:30 p.m. Thursday to Saturday; Complex 11 a.m. to 5 p.m. Friday and Saturday. Early American artifacts in several buildings. **Fee**: Donations are accepted.

Brookside County Park, 59 Brook St., Sayville, 631-563-7716. **Hours**: Open 1 to 3 p.m. Wednesday and Sundays year-round, weather permitting. This 5.8-acre "water park" on what was the former estate of architect Isaac H. Green features guided nature walks provided by the Great South Bay Audubon Society.

Edgewood Oak Brush Plain Preserve, off Commack Road in Central Islip, 631-444-0273. **Hours**: Year-round access by Department of Environmental Conservation permit (free). This preserve offers 650 acres of pine barrens and 50 acres of open field with a five-mile hiking trail popular with bird-watchers.

For a sampling of nearby restaurants, see Bayport-Patchogue area on Page 280 and Sayville on Page 293.

MONTAUK POINT LIGHTHOUSE

Watch Your Step, And Enjoy the View

ORMER ISRAELI Prime Minister Benjamin Netanyahu climbed it on his visit. As first lady, Hillary Rodham Clinton didn't have time. But each year, more than 100,000 tourists from far and near scale the 137 tower steps at **Montauk Point Lighthouse**, the equivalent of a spiraling 10-story trudge.

The payoff is Long Island's million-dollar view.

To the north, the clear day panorama embraces coastal Connecticut and Rhode Island; to the west, the windswept expanse of beach unfurling toward the Hamptons. Across open ocean to the south, a romantic optimist might conjure the Bahamas. And off to the east — could that be Portugal?

Through wars and hurricanes, pelting sleet and clam bisque fog, the tower has alerted ships to the shoals off Montauk Point ever since the original lantern was lit in 1797 (four months after the building was completed, because the ship bringing whale oil to fuel the beacon ran aground nearby).

New York's first lighthouse (and America's fourth-oldest among those still active), it was automated in 1987 by the Coast Guard, which continues to maintain the light and fog signal. But it's now owned by the **Montauk Historical Society** and open to the public as a museum.

Among the treasures displayed there is President George Washington's decree commissioning the lighthouse in 1792 — not coincidentally the same

A view of the 94.5-foot Montauk Point Lighthouse from the north side

year the New York Stock Exchange was established. After all, how could Treasury Secretary **Alexander Hamilton** turn his adopted city into a world financial center if every merchant ship met the same fate as the wrecked whale oil carrier? Even the arduous job of lighthouse keeper (which until 1961, when electricity finally came, included hand-cranking the lamp mechanism every three hours) was considered so vital that it, too, was bestowed by presidential appointment.

Exhibits depict the keepers' lonely work and document changes to the lighthouse over the decades. For example, the original 80-foot tower has grown to 94.5 feet to accommodate larger lenses. And about 1900, a reddish-brown band was added midway around the white cone; distinctive paint jobs, called "day marks," evolved to give ships a quick sight aid in distinguishing lighthouses from one another. No two lighthouses have the same flashing pattern, either (Montauk's is every five seconds).

Old photos capture the period in 1898, at the end of the Spanish-American War, when Theodore Roosevelt and his more than 23,000 Rough Riders were quarantined at Montauk's temporary tent village called **Camp Wikoff** (Roosevelt signed the lighthouse guest book while billeted at nearby Third House, but didn't note whether he'd climbed the tower).

The lighthouse's own battle has been against erosion, which in 200 years has nibbled almost 250 feet off its **Turtle Hill** promontory — bringing the bluff's edge to within 50 feet. But the bank has been stabilized for more than a decade thanks to **Long Islander Giorgina Reid**, who succeeded where the U.S. Army Corps of Engineers failed. Her solution: Terrace the bank with vegetation, which had done the trick at her North Shore summer home. For close to 20 years, starting on Earth Day in 1970, she drove to Montauk every weekend to direct the project (which after a few hearty laughs the Coast Guard approved). The museum's "Erosion Room" is dedicated to Reid, who died in 2001. Her patented method is chronicled in a book wryly titled "How to Hold Up a Bank."

On the lighthouse grounds are memo-

A few of the 137 steps on the spiral staircase inside the lighthouse

The automated light at the top of the tower

Newsday Photos / Ken Spencer

AT A GLANCE

Montauk Point Lighthouse is within Montauk Point State Park but operated separately; 631-668-2544, www.montauklight house.com. **Hours**: 10:30 a.m. to 6 p.m. most days from mid-June to early September (open later some days of holiday weekends); limited hours autumn through early spring; 10:30 a.m. to 5 p.m. daily from mid-May to mid-June (and on weekends through late October). **Fee**: $5 adults, $4.50 ages 62 and above, $2.50 ages 6 to 12 (children under 41 inches tall are admitted free but aren't allowed in the tower). Parking is $5 between 8 a.m. and 4 p.m. daily Memorial Day to Labor Day, weekends only most of the rest of the year. First-floor wheelchair accessible. Partially child appropriate.

Did You Know?
Family members, local American Indians and animal herders were early lighthouse keepers' main company, but in summer they could take paying guests.

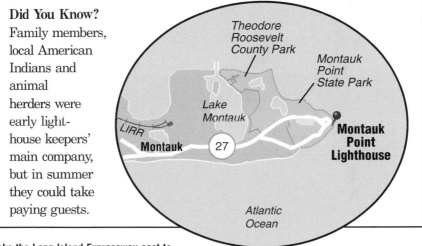

Take the Long Island Expressway east to Exit 70. Go south on Route 111 to Route 27 east. Follow Route 27 east nearly 40 miles to Montauk Point.

rials to Long Island commercial fishermen lost at sea and to the determined souls aboard the slave ship **Amistad**, who landed it at Montauk's **Culloden Point** in 1839 in a bid for freedom — which they won two years later after former President John Quincy Adams argued their case before the Supreme Court. There are occasional special events. And the coin-operated telescopes might help you clarify which Bahamas island you thought you saw from the tower's watch deck. ◆

Newsday Photo / Michael E. Ach

A restaurant overlooks Lake Montauk at Gosman's Dock, West Lake Drive.

Easygoing at End of the Road

INFINITELY FARTHER from the Hamptons in temperament than in driving time, low-key Montauk has the easygoing aura of end-of-the-road resorts from Key West, Fla., to Homer, Alaska. One of the few indications it's even aware of its just-folks allure is the droll sign that's graced the window of Shagwong Restaurant about as long as anyone remembers: "Piano player wanted. Must have knowledge of opening clams."

Montauk would be a very different place if the 1929 stock market crash hadn't cut short developer **Carl Fisher'**s plan to make it **Miami Beach North** (remnants include the Tudor-style Montauk Manor condo resort and a red-brick office tower jutting above the low-rise business district like a sawed-off sequoia).

Overall, the community remains a happy melange of fishermen's shanties, family friendly (and often family run) lodgings plus a few tony addresses such as Montauk Yacht Club and Gurney's Inn Resort & Spa.

Party central is a few motel-crammed blocks between the Atlantic Ocean and Main Street. Several miles north, along the western side of vast **Lake Montauk** where it opens into Long Island Sound (via a breach punched by the U.S. Army Corps of Engineers to create a safe haven for the fishing fleet after the 1938 hurricane decimated the old port), is the settlement of Montauk Harbor. It's a laid-back jumble of charter / party fishing / excursion boats, plus every type of eatery from clam shacks to those that actually ban bare feet and define sauce as something more than plastic packets of ketchup. The tourist center there is **Gosman's Dock**, started years ago as a boater's pit stop and now a seasonal waterfront mall of sorts.

Montauk is revered for its sportfishing and its ocean beaches — which stretch east to Montauk Point and west along roller-coastering Old Montauk Highway to **Hither Hills State Park.** But it offers a dollop of history, too. **Second** and **Third House Museums** were

Newsday Photo / Tony Jerome

A solitary fisherman casts for striped bass from the south side of Montauk Point, near Camp Hero.

18th- and early 19th-century settlers' homes (First House apparently wasn't rebuilt after it burned down). **Indian Field** burial ground, off East Lake Drive, is the last piece of Long Island owned by the Montauketts. Offseason? **Deep Hollow Ranch,** which bills itself as America's oldest cattle ranch (the nation's first cowboys were riding and roping on the site in the mid-1600s), offers beach / trail rides year-round. Hikers who need a focus other than Montauk's crashing surf can take part in winter seal walks. The seals, of course — like other true escapists — are always there. ◆

Photo by Doug Kutnz

A campfire and a sing-along at Hither Hills State Park

WHILE YOU'RE THERE

Montauk Point State Park, Montauk Point State Parkway, Route 27 east to end, 631-668-5000. **Hours**: Open sunrise to sunset year-round. **Fee**: $5 per car daily late May to early October, then weekends and holidays only. This 724-acre facility offers hiking, fishing, picnicking, cross-country skiing, hunting in season.

Montauk Downs State Park, off Southview Avenue, north of Route 27, 631-668-5000. **Hours**: Open sunrise to sunset year-round. **Fee**: Park entrance is free; call for tennis, golf and swimming fees. This 160-acre park offers an Olympic-size outdoor swimming pool and tennis courts open in summer, restaurants, and a championship 18-hole golf course designed by Robert Trent Jones open year-round (weather permitting).

Hither Hills State Park, off Route 27, four miles west of Montauk, 631-668-2554. **Hours**: Open sunrise to sunset year-round (campsites available early April to mid-November). **Fee**: $7 per car daily from late May to early September; no fee at other times. Besides the camping, this 1,755-acre facility offers fishing (including at night with permit), swimming, picnicking, playing fields, recreation programs, cross-country skiing, hunting in season; the "walking dunes" of Napeague Harbor are at the park's eastern boundary.

Theodore Roosevelt County Park, off Montauk highway, three miles east of the village, 631-852-7878. **Hours**: Open year-round. **Fee**: Call for information on fees and licenses. This park features hiking, biking, fishing, canoeing, hunting, horseback riding, outer-beach camping (no tents), picnicking, outer beach access as well as historic **Third House** (where Roosevelt stayed) featuring photos and artifacts of early Montauk plus a Spanish-American War exhibit. **Hours**: House open 8 a.m. to 5 p.m. daily from early May to mid-September; war museum is open 10 a.m. to 5 p.m. Wednesday through Sunday Memorial Day weekend to Labor Day. **Fee**: none.

Second House Museum, on the west side of the village, 631-668-5340. **Hours**: Open 10 a.m. to 4 p.m. daily except Wednesday in summer. **Fee**: $2, $1 younger than age 12. This 18th century farmhouse, the second house built in Montauk, was used by early herders.

For a sampling of Montauk restaurants, see Page 289.

MORTON WILDLIFE REFUGE
AND MASHOMACK PRESERVE

Where Birds Of a Feather Say Hello

T HE CLARION call "chicka-dee-dee-dee" ricochets around the tangled woods, and soon trailside boughs are dotted with portly little birds. Gently as falling leaves, one after another flutters down to delicately peck an offering of seeds from a visitor's outstretched palm. The black-capped chickadees of **Morton National Wildlife Refuge** have stolen another heart.

Since it became a nature preserve in 1954, feeding the unusually tame chickadees at the 187-acre Morton refuge in Noyack has been a treasured, if somewhat controversial, tradition. While not officially sanctioned, it is unofficially tolerated. "We don't recommend feeding wildlife," said a staffer, "but we know the chickadees are an attraction and we don't see it causing any great harm if it also has the benefit of making young children think about wildlife."

The birds certainly seem to enjoy the ritual (started, perhaps, by **Elizabeth A. Morton**, who bequeathed the land). Sometimes they even light on an empty hand despite an abundance of wild berries (and discarded store-bought seeds) along the grassy paths. (Unsalted, unshelled sunflower seeds are the recommended fare, but even these may not attract large flocks in summer, when the birds are busy catching more nutritious insects

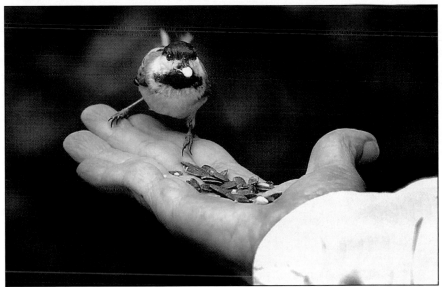

Newsday Photo / J. Michael Dombroski

At the Morton National Wildlife Refuge, the birds get friendly.

to bulk up their new hatchlings).

While the numerous chickadees have come to view humans as their friends — all right, as an easy food source — the preserve strives to keep visitors at a distance from its endangered piping plovers. The 1.5-mile **Jessups Neck** peninsula trail is off-limits during the April through August nesting season. To encourage respect for the beach closure, a sign details the birds' plight: "Development destroys their habitat; predators eat eggs and chicks; hikers and vehicles crush their nests and young; storm tides wash out their nests; pets harass them; people destroy breeding adults, which exposes eggs to the hot sun."

At the much larger **Mashomack Preserve** on Shelter Island, just a four-minute ferry ride from nearby North Haven, shore birds are protected by design: The 20 miles of trails are all inland. But since the preserve's 2,039 acres — nearly a third of Shelter Island — are edged by 10 miles of coastline, there are many water views. And the new **Coecles Harbor Marine Water Trail** is a five-mile round-trip route for canoers and kayakers; a map-brochure suggests 13 landing sites.

While Mashomack visitors can't expect close encounters with its residents — including chickadees — sharp-eyed hikers might spy these and other birds such as pheasants, ospreys and red-tailed hawks, as well as animals from turtles and salamanders to deer. The preserve — pronounced

Mash-AW-mack — belongs to the nonprofit **Nature Conservancy**, which calls it "a museum of life in process, a sanctuary where the natural cycles of flora, fauna and ecosystems are allowed to complete themselves free from human interference." Hikers thus are asked to look and listen but not jog, pick anything or stray from marked trails.

But there's plenty to see along the four main interconnected, color-coded loop footpaths. You can complete the 1.5-mile **Red Trail** in 45 minutes or tackle all four circuits for an 11-mile, four-hour hike through woods and fields. Though there are some low hills, none of the walks is strenuous.

There's also a barrier-free trail, which is a one-eighth-mile loop with interpretive plaques in large type and Braille winding past a native-plant garden and a freshwater "kettlehole." A one-mile hard-surface trail also is suitable for wheelchairs.

At the small visitor center, which has nature exhibits and a gift shop, you can borrow a map-brochure that expands on introductory information you'll see signposted at 18 spots along the Red Trail. They point out some of the preserve's more abundant and / or "interesting" highlights. Station No. 1: poison ivy. ◆

Newsday Photo / J. Michael Dombroski

A gazebo provides many visual angles at the Mashomack Preserve.

AT A GLANCE

Morton National Wildlife Refuge, Noyack Road (Route 38), 631-286-0485. **Hours**: Daily a half hour before sunrise to a half hour after sunset. **Fee:** $4 per vehicle. **Mashomack Preserve**, Route 114 (a mile from South Ferry), Shelter Island, 631-749-1001. **Hours**: Daily July and August, Saturday and Sunday in January, Wednesday to Monday other months; 9 a.m. to 4 p.m. in Eastern Standard Time, 9 a.m to 5 p.m. in Daylight Savings Time. **Fee:** Suggested donation $4 adult, $2 child. Both preserves child appropriate and partially wheelchair accessible.

___

Did You Know?
Shelter Island's first European settlers were Barbados sugar merchants seeking white oak trees for barrels.

From Sag Harbor, take Noyack Road west about two miles to the Morton Refuge. *For Mashomack,* take Route 114 north from Sag Harbor, drive through North Haven to the Shelter Island Ferry. From the ferry continue north on Route 114 (Ferry Road) for about one mile. Enter on the right. *From Greenport,* take the ferry to Shelter Island and follow Route 114 south. Take Route 114 about three miles to Mashomack on the left.

◆

Shelter Island Delights

B ICYCLISTS LOVE Shelter Island for its laid-back New England atmosphere and gently rolling, tree-shaded byways — many ending at vest-pocket beaches. To avoid the long weekend ferry lines, savvy day-trippers leave their car in **North Haven** or **Greenport**, walk their bikes aboard the boat (or rent them on the island), then meander around on two wheels.

The 8,000-acre island — whose irregular shape one resident described as a Rorschach inkblot gone mad — is even more compact than its official dozen square miles when you eliminate marshes and other areas not suitable for bicycling (including the huge **Mashomack Preserve**, which does, however, provide bike racks).

Sunset viewing is a major evening activity and there aren't too many planned daytime events. There's a scattering of shops, galleries and places to stop for a meal, snack or picnic fixings. But the overall snail's pace is just fine with the locals — and with visitors drawn by the island's simple charms.

Your itinerary should include lovely **Dering Harbor** — New York's smallest incorporated village (at last count, 28 households on 200 acres) — and sprawling **Big Ram Island**, en route to which you'll see huge osprey nests on platforms the phone company built for them atop its poles along the causeway. You also can visit a centuries-old Quaker cemetery off North Ferry Road.

On summer weekends, visitors can tour a couple of historical society properties — the home of a member of

A monument in the center of Shelter Island honors residents who died in World War I.

the First Continental Congress and a museum with changing exhibits in an old chapel — both in The Center, which is one of the island's two main communities. The other, known by an equally straightforward name — The Heights — has the island's oldest public building. It's the **Chapel in the Grove**, noted for its marine mosaic windows. It was built in 1875 to 1876 for guests of a fashionable hotel that was the center of a religious community. The gingerbread-trimmed houses on the streets between the chapel and the harbor also were part of the complex, and were built in the form of a cross. Earlier history is still being probed and visitors occasionally are invited to view the archeological dig at a 250-acre private estate once the core of a 17th century plantation that covered the entire island, supplying provisions for two Barbados sugar

WHILE YOU'RE THERE

Two historical society sites on Route 114 are open late June to early September, 11 a.m. to 3 p.m Friday to Sunday. The 1743 **Havens House,** 16 Southbury Rd., Shelter Island, 631-749-0025. **Fee:** $2 donation. The house, on the National Register of Historic Places, was built by William Havens, a member of the First Continental Congress. It contains mid-19th century additions with furnishings from the 18th and 19th centuries. The barn has farming and fishing artifacts. The 1890 **Manhanset Chapel,** 24 Northbury Rd., Shelter Island, 631-749-3429. **Fee:** $2 donation. This building originally stood in the village of Dering Harbor, where it served guests of an upscale hotel, and now has changing museum exhibits and hands-on activities for children.

The **Perlman Music Program's** camp for gifted students offers free seasonal performances at its summer home on Shore Road at Crescent Beach.

For a sampling of nearby restaurants, see Greenport-Cutchogue on Page 286, Sag Harbor on Page 292 and Shelter Island on Page 294.

Newsday Photos / J. Michael Dombroski

plantations. Last stop: **Crescent Beach** — for a soul-melting sunset and, in season, inspiring free concerts by gifted students at the summer home of the **Perlman Music Program** (co-founded by Toby Perlman, whose husband, violinist Itzhak Perlman, provides mentoring). ◆

Information: 800-974-3583, www.shelter-island.org; Perlman Music Program, 631-749-0740.

NORTH FORK WINE COUNTRY

Sweet Sipping On The Vineyard Trail

AT AN ART-FILLED chateau amid a sea of vineyards, a trio plays mellow jazz as fledgling oenophiles intently swirl, sniff, sip (and sometimes spit) — trying to fathom the difference between claret and merlot.

The scene could easily be in Bordeaux or Tuscany, but this is a totally Long Island experience — albeit one of relatively recent vintage. The chateaux are 20th or 21st century rather than 16th, and their owners not 10th-generation vintners but an assortment of artists, doctors and other entrepreneurs (including one movie mogul, one former Ivy League professor and one genuine Italian prince). The wines they produce on the North Fork already have gained worldwide acclaim.

But best of all, **"NoFo"** — as the area has been trendily dubbed — even outshines Europe's venerable wine regions in one key respect: It's little more than an hour's drive from home.

By filling their tasting rooms with art, music, informative programs and food-and-wine events in the tradition of more seasoned vineyards in Northern California's Napa and Sonoma valleys, Long Island's stylish wineries have turned the **East End** into a year-round tourist destination. Since many are family affairs, they feature a variety of child-friendly seasonal activities (and all stock water or juice for visiting kids as well as designated drivers). Also,

Ralph Pugliese, owner of Pugliese Vineyards in Cutchogue, shows off some chardonnay grapes on the vine.

their shops sell not only grape products, such as jams and jellies, but imaginative and well-priced themed gifts from crystal goblets to the inevitable T-shirt asserting **"Life Is a Cabernet."**

Sunny NoFo's ideal growing conditions (better than the foggier South Fork's) explain why wineries are popping up like champagne corks — in converted potato barns as well as distinctive new buildings designed to maintain the area's rustic character. The historic villages and low-key beach communities remain. It's just that now the main crop at many farms is grapes — some 2,000 acres of them all told since the first planting in 1973 by industry-founding Hargrave Vineyard (which became **Castello di Borghese-Hargrave Vineyard** in 1999).

At last count, 20 vineyards — from Aquebogue's **Paumanok** (its name the early Indians' word for Long Island) to Greenport's **Ternhaven Cellars** ("Last Winery Before France") — were open to the public along the roughly 20-mile North Fork Wine Trail. Most are on — or just a quick jog off — the two parallel east-west arteries: Routes 25 and 48. But the wineries are as different as their award-winning wines — which have been served from the White House to China.

As well as daily tastings, many offer frequent guided tours of the wine-making process. **Palmer** has created an interesting self-guided tour, complete with a trivia quiz. Others offer at least a picture-window view of stacked barrels or fermentation tanks. As well as indoor tasting rooms, most also have outdoor decks or terraces overlooking fields of vines. The typical profusions of flowers today are simply decorative, but at one time in Bordeaux, fragile roses were planted at the head of each row to warn of advancing root-killing blights. One winery, **Galluccio-Gristina**, invites visitors to walk into its vineyard, but most people reportedly take a few steps, sample one bitter-skinned grape and quickly retreat to the tasting room.

The price of popularity is that many wineries charge visitors a few dollars to taste each vintage (refundable with a purchase). Fair enough, most say.

Palate ready? For the most accurate assessment, sample light wines before heavy, white before red, and dry before sweet. By the way, you won't see an Old West spittoon on the floor, but most tasting rooms have a more refined looking — if still inelegantly named — "dump vase" on the bar for purists who don't want to consume all the wine they swish over their taste buds. Most visitors, however, hesitate to waste a drop. ◆

Newsday Photo / Ken Spencer

Cabernet sauvignon grapes at Palmer Vineyards

AT A GLANCE

All telephone numbers carry the 631 area code.
1. **Palmer Vineyards**, Aquebogue, 722-9463
2. **Paumanok Vineyards**, Aquebogue, 722-8800
3. **Jamesport Vineyards**, Jamesport, 722-5256
4. **Martha Clara Vineyards**, Riverhead, 298-0075
5. **Macari Vineyards**, Mattituck, 298-0100
6. **Lieb Family Cellars**, Mattituck, 734-1100
7. **Laurel Lake Vineyards**, Laurel, 298-1420
8. **Pellegrini Vineyards**, Cutchogue, 734-4111
9. **Galluccio Estate Vineyards-Gristina Winery**, Cutchogue, 734-7089
10. **Castello di Borghese-Hargrave Vineyard**, Cutchogue, 734-5111.
11. **Bidwell Vineyards**, Cutchogue, 734-5200
12. **Peconic Bay Winery**, Cutchogue, 734-7361
13. **Pugliese Vineyards**, Cutchogue, 734-4057
14. **Bedell Cellars**, Cutchogue, 734-7537
15. **Pindar Vineyards**, Peconic, 734-6200
16. **The Lenz Winery**, Peconic, 734-6010
17. **Raphael**, Peconic, 765-1100
18. **Osprey's Dominion Winery**, Peconic, 765-6188
19. **Corey Creek Vineyards**, Southold, 765-4168
20. **Ternhaven Cellars**, Greenport, 477-8737

For more information, please see Pages 298 to 300.

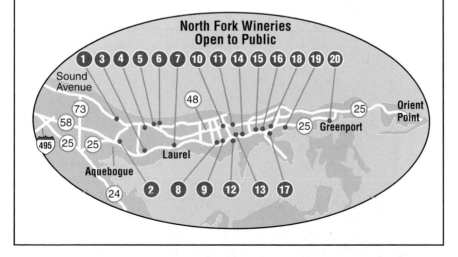

Take the Long Island Expressway east to Exit 73 (last exit). Follow Route 58 east until it merges with Route 25 in Aquebogue. Continue on Route 25 about one mile. Paumanok, one of the first of the North Fork's 20 wineries that is open to visitors, is on the left just past Church Lane.

Grapes are picked by machine and by hand. Above, a scene at Castello di Borghese-Hargrave Vineyard in Cutchogue and, below, Macari Vineyards in Mattituck

Picking, Fermenting, Enjoying

NO, YOU WON'T SEE women with purple feet stomping around in vats of grapes — that part of the wine-making procedure has been mechanized. But while the season and day of your visit somewhat determine what you *do* see on a winery tour, you're guaranteed fascinating glimpses of an age-old process unchanged except for a few equipment advances here and there.

The following tidbits were gathered on one of the free daily tours offered year-round at Peconic's **Pindar Vineyards**, Long Island's largest wine producer (about 80,000 cases annually) as well as its second oldest (Pindar's first vintage appeared in 1983).

It all naturally starts with the grapes. Most now are harvested by machines that shake them off vines to minimize damage (even though they're bound for the crusher). Grapes for premium wines are picked by hand.

Why do rows of grapevines often look like military platoons with identical buzz cuts? Annual pruning limits how much fruit is produced (typically five to eight pounds per vine, with each 2.5 pounds yielding one bottle of wine. "The object is to grow good grapes, not as many as you can," the tour guide said.

Newsday Photo / John H. Cornell Jr.

The East Hampton Sunshine Club tours Pindar Vineyards in Peconic.

Newsday Photo / J. Michael Dombroski

Visitors enjoy a sip and a view at Corey Creek Vineyards in Southold.

Newsday Photos / Tony Jerome

At Pugliese Vineyards, grapes are readied for the destemming machine. Below, wine is bottled at Bedell Cellars.

Harvested grapes go first into a destemmer-crusher, which smooshes them 10 tons at a time, producing 1,000 gallons of pulpy juice per batch. Next stop: the chilly stainless-steel fermentation tanks, where the sugar in the fruit is broken down into alcohol — taking about 10 days for 10,000 gallons. Skins stay with red wine through this process — that's what gives it the color. A month after the grapes are picked, you could have a bottle of light **"nouveau"** wine (with a shelf life of only several months). But most whites need to age in the tanks at least a year, reds two years. To develop more flavor, some wine varieties are aged in oak barrels for two years, then in their bottles another year before they're shipped.

Vines can yield wines with "nice complexity" five or six years after they're planted and can continue producing for a century. But like mature humans, they slow down as they get older — one of many factors that can influence a wine's price. And wine doesn't necessarily last forever. Champagne — the type most likely to be stashed in the refrigerator for a future occasion — won't go bad, but will go flat over time. The advice: "Drink it. What are you waiting for?" ◆

For a sampling of nearby restaurants, see Aquebogue-Southold on Page 280 and Greenport-Cutchogue on Page 286.

OLD BETHPAGE VILLAGE RESTORATION

Homes and Shops As a Patchwork Of Our Past

L AUGHTER SPILLS OUT of the Old Bethpage Village inn, circles the school yard in search of new allies, then fades into the buzz of daily commerce. This is a town that time *remembered*. But while it's now filled with life, it never actually existed as a community at all.

This makes perfect sense once you know the background of this booming burg — formally called **Old Bethpage Village Restoration**. Its several dozen buildings were moved here from around Long Island to create a typical mid-19th century rural farm settlement.

There's the former East Meadow inn (whose bar now serves root beer) and two East Norwich general stores (where 1800s purchases were routinely made on credit, though not credit cards). Barns have come from Syosset, Jericho and Old Westbury. Homes of simple tradesmen and prosperous merchants were plucked off crumbling foundations from Hempstead to Northville. Some 60 historical buildings so far (the village continues to grow) make up the 100-acre composite community on the Nassau-Suffolk border. Each was selected for its architectural detail and historic significance, and carefully restored to a specific date.

Adjoining this open-air museum is the site of the annual **Long Island Fair** — replicating the original 1866 Queens County Agricultural Soci-

Jugglers perform at an annual fall fair at Old Bethpage Village Restoration.

Newsday Photo / Michael E. Ach

Costumed interpreters chat among the many preserved buildings at Old Bethpage. Below, children compete in an October sack race.

Newsday Photo / Tony Jerome

AT A GLANCE

Old Bethpage Village Restoration, Round Swamp Road, Old Bethpage, 516-572-8400. **Hours**: Wednesday to Sunday March through December; hours vary by day and season. **Fee**: $6 adults, $4 ages 5 to 12 and over 60. Partially wheelchair accessible. Child appropriate.

Did You Know? Boys usually attended school less than girls in mid-1800s rural communities because they had to help with harvesting and other farm chores.

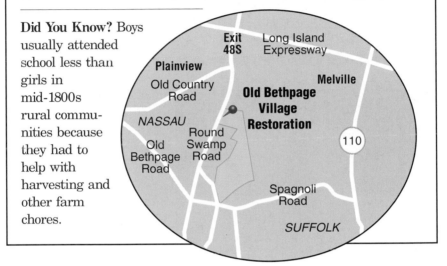

Take the Long Island Expressway to Exit 48S. Drive south about one mile on Round Swamp Road. Old Bethpage Village Restoration is on the left.

ety Fairgrounds, which had been located in Mineola.

The entire 210-acre complex occupies land farmed by Richard S. Powell in the mid-1800s, and his original home and carriage shed are still central to a working farm that includes cows, sheep, pigs, horses and oxen (don't be surprised if you bump into a few sheep wandering around the nearby Williams Farm, too).

Costumed interpreters blow the dust off history's dry details and bring the period to life. In the **Benjamin House**, a bonneted docent points out the tiny "pauper's room" — explaining how families submitted bids to the town "overseer of the poor" for the amount they'd need to take in a lost soul (in one 1829 case, the accepted low bid was 65 cents a week plus clothing).

At the one-room schoolhouse (formerly a fixture in Manhasset) another docent introduces visiting students to the rigors of education back when teachers had to fire up the wood stove before starting lessons, and mischief-makers were swiftly disciplined at home by stern fathers tipped off by tattling siblings.

The simple **Manetto Hill Church**, founded in Plainview during a burst of mid-1800s Methodist activity, was at the time one of several served by a circuit minister. Today, it is still witnessing weddings, vow renewals, baptisms (and once a bar mitzvah). On Memorial Day, a procession proceeds from the church to a ceremony at the village cemetery (whose tombstones were reclaimed from abandoned sites and no longer mark real graves).

Many Old Bethpage buildings were both residence and workplace for their owners, and the **1866 Layton Store & House** portrays a thriving business and a lifestyle to match. Gadgets and souvenirs are sold across the crossroads at the **Luyster Store** — formerly a North Shore landmark for generations of locals including Theodore Roosevelt. (Both shops stood in East Norwich near the intersection of Routes 25A and 106.)

Newsday Photo / Michael E. Ach

A player for the Glen Head Zig Zags bats during an old-time game at Old Bethpage.

Another bygone celeb is linked to the **1820 Conklin House** (moved from the Village of the Branch). Poet Walt Whitman supposedly lived there for a year or two while he was teaching school in Smithtown.

Every season also sparks special events at Old Bethpage — all naturally delivered in authentic old-fashioned style. The **Spring Festival** includes crafts and horticultural exhibits. Summer brings 19th century band concerts, baseball games and a rousing July Fourth celebration. Fall signals the Long Island Fair, featuring more fun and entertainment — plus prize-winning foods, produce and livestock. December heralds candlelight and caroling — followed by a brief mid-winter rest (when 1800s villagers undoubtedly would have loved to fly down to Florida). ◆

Newsday Photo / Karen Wiles Stabile

A Grumman Avenger torpedo bomber is guided onto the Republic Airport runway at the American Airpower Museum in Farmingdale.

A Short Flight Into the Past

A GRUMMAN AVENGER whose fold-up wings saved space aboard aircraft carriers, and one of the few remaining airworthy P-47 Thunderbolts built on the site when it was home to Republic Aviation, are star members of a flight-ready World War II squadron at the **American Airpower Museum** at Republic Airport in Farmingdale.

When they're not being put through their paces in the air, the restored vintage aircraft — many personalized by their wartime pilots with distinctive "nose art" — are displayed in a hangar that was part of the Thunderbolt assembly line when **Republic** and Bethpage's **Grumman** were central to Long Island's defense industry. Republic's runways dispatched some 9,000 of these torpedo bombers (former President George H.W. Bush was shot down over the Pacific in such a plane).

Kids can climb into gun turrets, cockpits and deactivated ejection seats, and check out a re-created frontline Briefing Room and Ready Room (where even adults will struggle to heft a weighty flak jacket).

You can also inspect a variety of uniforms and search for familiar faces

One of many warplanes on exhibit at the American Airpower Museum

among the war photos (don't miss actors Clark Gable and James Stewart). Dioramas depict battle scenes and there's a model of the **USS Arizona Memorial** in Honolulu (commemorating the Japanese attack on Pearl Harbor).

A special Hangar 3 exhibit chronicles the problems and contributions of the **Tuskegee Airmen**, a squadron of black fighter pilots who first had to fight for a chance to join the U.S. Army Air Corps.

Republic's original 1943 control tower across the tarmac from Hangar 3 is being restored to its 1940s heyday. It has a mock-up officer's club, more uniforms and, on the top floor, the original communications equipment — looking as antiquated as the World War II planes, but like most of them still in working order.

Volunteers giving the guided tours can all offer personal perspectives of their own wartime exploits — many of which occurred in the air.

Want to outfit yourself in similar fashion? The museum gift shop carries classic bomber jackets, hats, patches and emblems, plus books and aircraft models. ◆

American Airpower Museum, 1300 New Highway, Farmingdale, 631-293-6398. Hours: Open Thursday to Sunday 10:30 a.m. to 4 p.m. Fee: $9 adults, $6 seniors and veterans, $4 ages 5 to 13.

WHILE YOU'RE THERE

The **Hicksville Gregory Museum**, Heitz Place and Bay Avenue, Hicksville, 516-822-7505 and www.gregorymuseum.org. **Hours**: 9:30 a.m. to 4:30 p.m. Tuesday to Friday, 1 to 5 p.m. weekends. **Fee**: $5 adults, $3 ages 6 to 16 and seniors. The museum, housed in an 1895 former courthouse (complete with old jail cells), contains Long Island's largest rock and mineral collection (about 10,000 specimens, including fluorescent minerals that glow under ultraviolet light), fossils (including a rare pair of 100-million-year-old dinosaur eggs with baby dinosaur skeletons still intact), more than 5,000 butterfly and moth specimens from around the world, plus changing exhibits on local history.

Belmont Lake State Park, Sylvan Road, North Babylon, off Southern State Parkway, exit 38, 631-667-5055. **Hours**: Sunrise to sunset year-round. **Fee**: $5 per car daily in summer; $5 weekends and holidays spring and fall. This 459-acre former country estate of August Belmont offers biking, walking and bridle trails (horse rentals available at Babylon riding center, 631-587-7778), freshwater fishing with license from lakeshore or rowboat (boat rentals in season), three softball fields, picnic and play areas, and horseshoe pits.

For a sampling of Bethpage-Plainview restaurants, see Page 281.

Newsday Photo / Michael E. Ach

Children enjoy a church picnic at Belmont Lake State Park.

OLD WESTBURY GARDENS

A Breathtaking Eden In Full Bloom

WHEN AMERICAN steel fortune heir **John "Jay" Phipps** fell for globe-trotting shipping heiress **Margarita "Dita" Grace,** he promised her not only a rose garden but a country estate to rival her favorite family home — a formidable English landmark called **Battle Abbey,** built 900 years ago by William the Conqueror.

Phipps transformed an old Quaker farm into a Long Island eden now known as **Old Westbury Gardens** — living proof there's no limit to what a rich man (and his estate designer) can do for the woman he loves.

While palatial **Westbury House** is logistically the estate centerpiece, the antique- and art-filled mansion is eclipsed by the knockout grounds: about 160 acres of formal gardens, perennial borders, sweeping lawns and natural ponds with a "forever wild" perimeter that today still effectively screens the outside world.

As well as a rose garden (patterned after one at Battle Abbey, of course), the Phippses established a grove of specimen conifers known as a pinetum (common in England, where there aren't many native pines); a "Ghost Walk" of hemlocks echoing Battle Abbey's "Yew Walk," where ghostly monks are said to stroll on moonlit nights (reportedly no spirits inhabit Old Westbury's garden path); and an Italianate walled garden. Ornamental fountains

A visitor sketches a tranquil scene at Old Westbury Gardens.

Photo by Thomas A. Ferrara

Inside a walled garden on the Old Westbury grounds

and statuary dot the grounds, walks meander through tunnels of fragrant vines, and grassy tree-lined avenues, or "allees," stretch straight as an Iowa interstate for up to half a mile. There's even a rose-covered sandbox.

Sound like a movie set? It's been one many times — for example, for "The Age of Innocence" (a 1993 drama whose crew painstakingly tied silk blossoms onto dormant rose bushes for the offseason filming) and for Woody Allen's 2001 film "The Curse of the Jade Scorpion."

Since 1959 — about a year after Jay and Dita died (within months of each other, after more than 50 years of marriage) — the estate has been open to the public about half of each year for tours, lectures, plant sales and various entertainments from Christmas celebrations to a popular **Picnic Pops** summer concert series with dancing under the stars. Today's grounds also include a test rose garden and demonstration gardens you might adapt to your own quarter acre.

When Jay Phipps was first scouting locations for his estate, he zeroed in on the flat, mid-Nassau plains as perfect for one of his priorities: a polo field (which was completed in 1905, two years before the house). Not to be outdone, Dita Phipps went about recreating the landscape of her beloved

AT A GLANCE

Old Westbury Gardens, 71 Old Westbury Rd., Old Westbury; 516-333-0048; www.oldwestburygardens.org. **Hours**: 10 a.m. to 5 p.m. Wednesdays through Mondays, late April through late October; open Sundays in November until Thanksgiving. Special events during the December Holiday Celebration; call ahead for dates and times. **Fee**: Gardens only: adults $8, ages 62 and over $6, ages 6 to 12 $3 (house and gardens $10, $8 and $6). Wheelchair accessible. Children under 16 must be accompanied by an adult.

Did You Know?
Old Westbury Gardens boasts Long Island's largest tree peony (5 to 6 feet tall) and some of its oldest beeches.

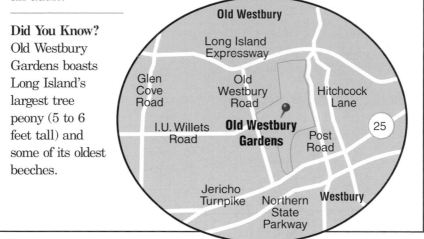

Take the Long Island Expressway to Exit 39S. Continue east on the service road, 1.2 miles on to the first road on the right, which is Old Westbury Road. Continue a quarter-mile to the entrance on the left for Old Westbury Gardens.

Sussex (the rose garden was being created along with the house).

Both the home and the gardens were designed by London "esthetic architect" George Crawley — who had exquisite taste but no formal training (to ensure that his buildings would stand up, he grudgingly worked in collaboration with a bona fide structural architect).

Mature trees were purchased from **Henry Hicks,** whose family nursery is still down the street, and somehow maneuvered into the ground. The hundreds of pairs of beeches and lindens along the three allees now tower over strollers like Brobdingnag palace guards.

Westbury House, completed in 1906. A Latin inscription at the door says, "Pax Introentibus — Salus Exeuntibus," or "Peace to those who enter. Good health to those who depart."

Although the Phipps family, which eventually included three sons and a daughter, wintered in Florida, Old Westbury clearly was considered home. Amid the gardens lies a sizable pet cemetery as well as a glorious toy-filled playhouse called the **Thatched Cottage**, which was the province of daughter Peggie (who in 2002 still lived on a neighboring estate). Today's young visitors can't play there, but they do have the run of her brothers' more basic log cabins.

During World War II, the estate also was a playground for 30 children of English friends and relatives the Phippses brought to Long Island to escape the **Battle of Britain**. The children boarded with various nearby family members but gathered at Old Westbury to share Sunday dinner and frolic on the grounds — where it seems love was destined to forever bloom. ◆

Photos by Thomas A. Ferrara

An entrance to the Walled Garden at Old Westbury Gardens

Inside Westbury House

ENGLISH ARCHITECTS. English nannies. English roof tiles and mantelpieces. Long Island would be Dita Phipps' home base throughout her adult life, but was there any doubt where she'd left her heart?

Westbury House is still adorned with family pictures and mementoes, and remains essentially unchanged from when they lived there. It was expanded and adapted as the family grew, however. (The present-day gift shop had been at various times a children's dining room, smoking room, billiards room, housekeeping room and part of the kitchen and butler's pantry.)

Decorative touches to take special note of as you tour the house include the carved plaster fruits and flowers on the ceiling in Jay Phipps' study (which was originally designed to be the dining room); the book-lined walls of Dita Phipps' study (one panel is just a facade concealing a niche where two priceless violins were secreted); intricately ornate mirrors in the Red Ballroom; Chinese Chippendale window decoration in the White

The White Drawing Room was a favorite of the Phipps family, especially during harsh weather.

The West Porch windows could be hydraulically lowered into the basement.

Drawing Room; and the master bath's wicker "throne." Oh — why did armoirs instead of closets serve most of the home's storage needs? You'll find out from a docent. (A clue: It had to do with tax regulations in — where else? — England.)

Designer **George Crawley**, who collaborated on Westbury House with architect Grosvenor Atterbury, also had a hand in the interiors of the Fifth Avenue town house of Jay's father, Henry Phipps. So when the Manhattan house was sold and destined for demolition, its dining room was moved lock, stock and trompe l'oeil ceiling (a realistic looking cloudy sky motif) to Westbury House.

But your favorite room (and one of the Phippses'), is likely to be the beamed ceilinged **West Porch**, whose massive windows all can be lowered hydraulically into the basement. It was on this all-season porch that Jay and Dita Phipps faithfully observed the tradition of — surprise, surprise — English afternoon tea. ◆

Photos by Thomas A. Ferrara

The West Porch from the outside

WHILE YOU'RE THERE

Clark Botanic Garden, 193 I.U. Willets Rd., Albertson; 516-484-8600. **Hours**: Open daily, 10 a.m. to 4 p.m. **Fee**: None, though donations are appreciated. This garden is an easily walked

12-acre "living museum" of more than 1,000 labeled varieties of trees, shrubs and plants chosen for their beauty, adaptability and availability. Curving sand paths. Now part of the Town of North Hempstead's Parks & Recreation Department, it's the former home of Grenville Clark, an adviser to President Franklin D. Roosevelt. Numerous public programs for adults and kids. Plant clinic, gift shop.

Newsday Photo / Karen Wiles Stabile

An egret spreads its wings on a July day at the Clark Botanic Garden in the Town of North Hempstead.

Goudreau Museum of Mathematics and Science, 999 Herricks Rd., Herricks Community Center, Room 202, New Hyde Park, 516-747-0777 or e-mail info@mathmuseum.org. **Hours**: Monday to Friday by appointment, year-round; noon to 3 p.m. most Saturdays, October to May. **Fee**: $2 Saturday; additional fee for workshops. This hands-on learning center, which has games and puzzles, offers demonstrations, workshops and special programs for kids 6 and older. Offers Saturday math enrichment for grades four to eight.

The Knothole, Christopher Morley Park, Roslyn, 516-571-8131. **Hours**: 1 to 5 p.m. Sundays, June to October. The restored writer's studio of Christopher Morley, author of "Kitty Foyle" and other novels; bathroom designed by Buckminster Fuller.

For a sampling of nearby restaurants, see Westbury on Page 296.

THE PINE BARRENS

Hiking The Near Wilderness

DAY PACKS bulging with trail mix, spare socks and enough bug spray to outfit an African safari, 18 hikers met at a rural crossroads one Saturday to explore a chunk of Long Island's largest natural area — the 100,000 acres of scrub trees, ponds and salt marshes known collectively as the **Pine Barrens**.

Stretching across Brookhaven, Riverhead and Southampton towns, it sits atop an aquifer that is Long Island's largest source of pure drinking water — a key reason conservationists fought hard for a development-free core area that's almost five times the size of Manhattan.

It includes named tracts such as **Sears Bellows County Park** and **Cranberry Bog County Preserve**, plus the entire length of the Peconic River. Most areas are open to independent hikers, but for companionship, convenience and security, many outdoor fans prefer to join a group. Each year, the **Long Island Greenbelt Trail Conference** schedules about 150 free, volunteer-led treks in the Pine Barrens (and along the Long Island and Nassau-Suffolk greenbelts). When the final 15-to-20-mile link in Southampton is completed, hikers will be able to trek 120 miles from Rocky Point to Mon-

On the Pine Barrens Trail, which extends from Rocky Point to the Hamptons

tauk along the main-drag Paumanok Path. But it already provides an alluring slice of near-wilderness.

"I live in Queens, in a building that has a gym right downstairs," said one hiker. "So why did I drive 50 miles out here? Because nature is the best stress buster there is."

That sentiment seemed representative of the group, a mostly middle-aged bunch of avid walkers about evenly divided between men and women, who had converged from across Long Island for what was billed as a moderate / flat hike of six to seven miles with views of Sears and Owl Ponds, **Flanders Bay** and an old cemetery. (Other scheduled hikes range from an "easy" two to four miles to a "moderate, very hilly" 22 miles.)

Guide-for-the-day Nancy Duffrin started with a caution about ticks and mosquitoes (listen up, or be sorry later), then set a brisk pace. Her circular route encompassing both sides of **Route** 24 in Flanders wound along marked and unmarked trails — sometimes narrow paths paved with leaves and pine needles requiring hikers to troop along in single file, other times on sandy access roads wide enough for four abreast.

Nancy and others offered snippets of local history and suburban hiking lore. But chatter was minimal. Everyone was there foremost to com-

The Pine Barrens Trail Information Center in Manorville

mune with nature — to listen to the chirping of birds and the whisper of wind in the pines (and perhaps to an inner muse).

Hikes out in the **Hamptons** have been known to attract an occasional dilettante who whines that she needs to get back for happy hour or asks where the bathrooms are. (The answer to the latter: This is the woods, babe — there usually are none; humans on most serious hikes must answer nature's call as the deer and squirrels do, albeit with a tad more modesty.)

A pinecone at Sears Bellows County Park

Sometimes the trail was bordered by cattails and marsh grasses, at others it traversed high and dry hillsides overgrown with berry bushes and shaded by scrub oaks and aromatic pitch pines, whose deep tap roots and rough bark full of corky air pockets help them survive near-desert conditions and fires as hot as 2,000 degrees. A huge osprey nest

AT A GLANCE

The Long Island Greenbelt Trail Conference office at Blydenburgh County Park, 631-360-0753 or www.hike-li.com. **Hours**: Open 10 a.m. to 4 p.m. Tuesday to Friday, noon to 4 p.m. Saturday (tours of historic Blydenburgh House are given Saturdays at 1 p.m.). The **Pine Barrens Trail Information Center**, 631-369-9768, a quarter-mile north of LIE Exit 70 in Manorville. **Hours**: Open 9 a.m. to 5 p.m. Friday to Monday Memorial Day through Columbus Day weekend. Partially wheelchair accessible. Child appropriate.

Did You Know? L.I.'s Pine Barrens is New York's third protected forest preserve, joining the Adirondacks and the Catskills in this status in 1995.

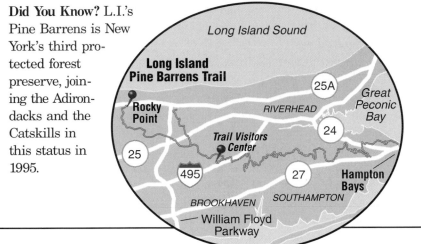

To reach the Pine Barrens Trail
Information Center, take the Long Island Expressway to
Exit 70 north. Turn left. The center is one-quarter mile north on the right side of the road.

looked down on **Mill Creek**. But there was no quicksand, no bottomless canyons and some bearberry but no bears.

Despite only a brief lunch stop — with plenty of offers to share extra sandwiches or snacks — no one ever lagged behind or complained of being tired during the four-hour trek. How could that be?

"You can relax in the woods because you can free your mind," said the philosophical hiker, who was hoping that particular day would help her cope with the tragic death of a friend the previous week. "Surrounded by nature," she said, "you can come to terms with the continuum of life." ◆

Tips for the Trails

L ONG ISLAND'S **Pine Barrens** is hardly the Sahara Desert, but it's remote enough to warrant precautions that might not be necessary in more manicured — and monitored — nature preserves.

Hikers should leave word with someone pinpointing the area they plan to explore and take maps, a compass, simple first-aid supplies, adequate food (fruit and lightweight edibles that needn't be refrigerated are good choices), plenty of water, extra socks, clothing layers that can be adjusted according to weather, plus a water-resistant shell. Wear comfortable walking shoes (heavy hiking boots usually aren't necessary).

Also recommended: Call the **Long Island Greenbelt Trail Conference** in advance to determine the current state of trails you plan to use. And note, hunting is allowed in various areas October through January.

The organization says that while **mosquitoes** are mainly a summer scourge, you can pick up **ticks** anytime.

Hikers can generally avoid them — and the risk of Lyme and other diseases transmitted by deer ticks — by taking these precautions:

Consider insect repellent (but avoid prolonged use).

Wear long pants tucked into socks and long-sleeved shirts tucked into pants.

Stay on trails and avoid grassy, brushy areas (even during pit stops). And avoid sitting on the ground — especially near ponds (frequented by deer) or in areas filled with acorns (which attract mice and other small rodents that often are infected by deer ticks).

Check your body thoroughly after a hike. Deer ticks, unlike dog ticks, are hard to see. In the first of three growth stages of their two-year life cycle, they're no larger than the period at the end of this sentence. Though easy to miss, this late-summer larval stage is the least dangerous because as newborns deer ticks haven't yet been exposed to infections. At the next, most dangerous, stage (which peaks late May through July) they're still no bigger than a poppy seed but now are looking for a second host. Adult ticks that have been on two hosts — and thus are twice as likely to have picked up diseases — actively seek new victims from September through late October-early November, then again in March and early April.

Should you find a tick attached to you, grab it as near to your skin as possible with tweezers. Pull it straight out without twisting. If you develop a rash or flu-like symptoms, consult a doctor. ◆

WHILE YOU'RE THERE

Newsday Photo / Tony Jerome

An old East Moriches truck at the firefighters museum in Ridge

Brookhaven Volunteer Firefighters Museum, Fireman's Memorial Park, Route 25 (Middle Country Road), Ridge, 631-924-8114, www.brookhavenfiremuseum.org. **Hours:** Open Tuesday and Thursday 10 a.m. to 3 p.m., Saturday 10 a.m. to 4 p.m., Sunday noon to 4 p.m. (Saturday only November through April, 10 a.m. to 2 p.m.). **Fee:** Donation appreciated. Restored 1889 Center Moriches firehouse with two floors of memorabilia. Six pieces of equipment in a truck house.

Old Schoolhouse Museum, 90 Quogue St., Quogue, 631-653-4111. **Hours:** Open July and August, 3 to 5 p.m. Wednesday and Friday, 10 a.m. to noon Saturday. **Fee:** None. The 1822 schoolhouse contains local memorabilia including photos, toys, farm equipment.

Quogue Wildlife Refuge, Old Country Road, Quogue, 631-653-4771. This 305-acre preserve has seven miles of self-guided trails open daily year-round, and a nature center open 1 to 4 p.m. Tuesday, Thursday, Saturday and Sunday.

The **Big Duck,** Route 24, Flanders (at the entrance to Sears Bellows County Park), 631-852-8292. **Hours:** Open 10 a.m. to 4:45 p.m. daily from May through September, then Friday through Sunday to mid-December. This unusual landmark, listed on the National Register of Historic Places, began its life as the retail store for The Big Duck Ranch in Riverhead, where it roosted from 1931 to 1936. The 20-foot-high tourist attraction, which originated a genre of sculptural roadside architecture, now houses a gift shop operated by the nonprofit Friends for Long Island's Heritage and also stocks some East End tourist literature.

For a sampling of nearby restaurants, see Hampton Bays-Yaphank area on Page 286 and Riverhead on Page 292.

PLANTING FIELDS ARBORETUM

A Regal Garden In Oyster Bay

THE GLORIOUS GARDENS and luxuriant lawns might have been peeled off a proper English estate and flown here magic carpet style — complete with the Tudor mansion.

It didn't quite happen that way, but the property that now is **Planting Fields Arboretum State Historic Park** *was* developed by a transplanted Englishman. He imported everything from the treasured 1711 Carshalton gates (which once guarded the estate of a former lord mayor of London) to a shipment of rare tropical camellias (which he naively stuck in the cold Long Island ground when they arrived in full bloom from the balmy Isle of Guernsey one February day — then had to quickly get a greenhouse built around them).

Hoping to smooth the path for other novice gardeners, he deeded the 409-acre Oyster Bay property to the state in 1949 with a proviso that it be put to educational use. And so the former home of insurance tycoon **William Robertson Coe** has become a living textbook for budding horticulturists as well as a stunning showcase of art, architecture and design. His legacy surely proves that the words "visionary" and "insurance executive" are not mutually exclusive.

The **Synoptic Garden** — a 5-acre synopsis of more than 400 species of

Rhododendrons line a shaded path at Planting Fields Arboretum.

flowering trees and shrubs arranged in alphabetical order from abetta to zenobla to make them easy to locate — is a perfect example of the arboretum's user-friendliness. And its half-acre of indoor plant space — larger than most suburban backyards — is a colorful refuge year-round.

Visitors can learn about everything from silk flower arranging to local archeology, and enjoy indoor and outdoor concerts run by **Friends of the Arts**. Kids can take in a variety of special performances on the annual family weekend during Planting Fields' December winter festival, and can climb trees on Arbor Day.

These fields were first used for planting by the Matinecock Indians, and Coe farmed them, too. The grounds — now about half cultivated — were laid out by noted landscape firms including the famous **Olmstead Brothers**. Even the greenhouses and outbuildings were planned by top designers of the early 20th century. The most significant of the property's 32 gardens were located near the grand mansion, now called **Coe Hall**.

Coe had a particular liking for new plant varieties and modern growing techniques, and he brought many spectacular specimen trees there

179

Flowers in bloom in January at the Camellia House at the former estate.

Fall crocus at the arboretum

AT A GLANCE

Planting Fields Arboretum State Historic Park, Planting Fields Road, Oyster Bay, 516-922-9200 and www.plantingfields.org. **Hours:** The grounds are open 9 a.m. to 5 p.m. daily except Christmas (greenhouses close at 4 p.m.). **Fee:** Admission is $5 per car from May through Labor Day, then weekends only. **Coe Hall** is open for guided tours. **Hours:** Noon to 3:30 p.m. daily (except holidays) April through September. **Fee:** $5 adults, $3.50 seniors, $2 ages 6 to 12. Grounds and house partially wheelchair accessible. Child appropriate.

Did You Know?
Like most Gold Coast estates, Planting Fields wasn't quite self-supporting, but the point was more to mirror the country manors of British royalty.

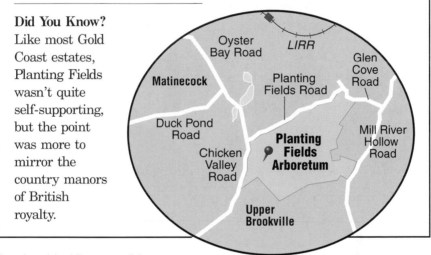

From Long Island Expressway Exit
41N, take Route 107 (Cedar Swamp Road) north about
three miles to Wolver Hollow Road. Turn right onto Wolver Hollow and go about three miles to
Chicken Valley Road. Make a right onto Chicken Valley and go about one mile. Arboretum is on right.

full grown to give the appearance of a mature landscape. In one of the largest-ever tree-moving operations in the Northeast (requiring the widening of a two-mile stretch of road and temporary removal of utility lines), two 60 foot purple beeches were barged to Oyster Bay from the Massachusetts childhood home of Coe's second wife, Mai (whose father was a founder of Standard Oil). One **"Fairhaven beech"** still stands on the west lawn, next to another signature tree, a weeping silver linden.

Outdoors, there's always something on display: lilacs, rhododendrons, magnolias, wildflowers, roses spring to fall, then dahlias. You certainly

Newsday Photo / J. Michael Dombroski

A bench beneath a cherry tree welcomes the weary at the arboretum. The fields were first used for planting by the Matinecock Indians.

don't need to go to New England for autumn leaf-peeping. And when the reds and golds turn to brown, and even the fruits and berries have dropped off, you can still admire bark textures and branch shapes as you wander the several miles of trails and paths shaded by oak, maple, beech, sassafras and black birch, as well as plenty of evergreens on the natural southwestern edge.

After most trees have shed their leaves — and you and the rhododendrons are curling your fingers and toes to keep warm — you can enter a tropical garden or showy Christmas display of poinsettias in the Main Greenhouse. And January through March, the **Camellia House** is filled with the red, white and pink blossoms of more than 300 of these rare natives of China and Japan. This is the Northeast's largest collection under glass — about a third of them survivors of Coe's original ill-timed batch from Guernsey. ◆

Newsday Photo / Ken Spencer

Ripe crab apples are a sure sign of autumn.

The Buffalo Room at Coe Hall is adorned with raised-figure murals that follow an Old West theme.

Olde English, Old West

WILLIAM AND MAI COE so loved their gardens that when the home they had acquired with the estate burned down, they made sure the replacement house exactly fit its footprint so as not to disturb the foundation plantings.

The design of the new 65-room limestone mansion (half-timbered on two sides to give the illusion it had been built over a period of centuries instead of three years) was inspired by several late 16th and early 17th century English manors. Personal touches include a storm-tossed ship over the front entrance, signifying Coe's marine insurance specialty (among his clients: the **"unsinkable" Titanic**). Carvings along the roofline south of the dining room include the faces of his valet, chauffeur and favorite hunting guide.

At **Coe Hall**, the "Olde English" interior style popular in the early 1900s features antique tapestries; stone and marble mantelpieces; carved ceilings, arches and paneling, and exquisite stained glass. The guest wing ech-

---

WHILE YOU'RE THERE

Other places of interest in the area include **Tiffany Creek Preserve** (Sandy Hill Road, Oyster Bay, 516-571-8500), **Raynham Hall** (20 W. Main St., Oyster Bay, 516-922-6808) and the Oyster Bay Historical Society's circa-1720 **Earle-Wightman House** (20 Summit St., Oyster Bay, 516-922-5032). For details on all, see Page 215.

Also **Muttontown Preserve** (Muttontown Lane, East Norwich, 516-571-8500) and **Charles T. Church Nature Sanctuary / Shu Swamp** (Frost Mill Road, Mill Neck). For details on these, see Page 115.

For a sampling of nearby restaurants, see Glen Cove on Page 284, Locust Valley on Page 288 and Oyster Bay on Page 290.

---

oes an English country village: the hall its main street, rooms the "shops" along a raised "sidewalk."

There are departures from Elizabethan England, too. A Dutch chandelier over the dining table celebrates the founding of an 18th century Sephardic congregation in Antwerp. The ornate **Louis XVI Reception Room**, like the half-timbered siding, was meant to give the house — and the family pedigree — the appearance of a long evolution. And Mai Coe's frilly boudoir reflects, well, her philosophy that you never can combine too many design elements. You won't believe this space was decorated by the same artist whose murals stampede across the walls of the so-called **Buffalo Room**.

Coe became enamored of the West on a 1905 hunting trip, and in 1910 bought William F. "Buffalo Bill" Cody's remote Wyoming ranch and commissioned the same architectural firm that designed Coe Hall to build a lodge there. The Coes visited every year by private railway car — accompanied by an entourage of servants — and in 1912 he made Wyoming his official residence (Could the taxes have been lower than Nassau County's?). Over the years, Coe made a number of gifts to the **Town of Cody**, including its first paved streets and medical center, and established American studies programs at 40 U.S. colleges. But he also enjoyed his garden-filled "springtime estate" in Oyster Bay until 1955, when he died in his newest abode, in Palm Beach. ◆

PORT JEFFERSON

A North Shore Port Of Many Callings

IF EVER a town needed a new image, it was **Drowned Meadow**, a marshy colonial shipbuilding center where shoppers had to don hip boots if they wanted to browse for scented candles and wind chimes at high tide.

Things improved in the mid-1800s, when America's third president helped fund a project to stop the business district's daily floods. To honor him, the village was renamed **Port Jefferson**. Tourism grew steadily over the next century, and has been the major industry by far since about the Mochaccino Era of the Italian-Ice Age.

Visitors now descend from all directions — many crossing Long Island Sound from New England via Bridgeport & Port Jefferson Steamboat Co. ferries. An early backer of the line was legendary 19th century showman **P. T. Barnum**, who'd planned to open a recreation center in nearby Poquott and make Port Jeff his traveling circus' winter home (when locals nixed these plans, he settled for developing the Brick Hill neighborhood, where the model he built is now known as the Barnum House). One ferry also is named for him, and the sight of today's tourist hordes pouring off the boat toward Port Jefferson's waterfront shops recalls his wry observation that, "Every crowd has a silver lining" (he's only alleged to have cracked the famously jaded, "There's a sucker born every minute").

Dining at Danford's Inn on Port Jefferson Harbor

Port Jeff's prices actually are as friendly as the small-town atmosphere, which is ever-welcoming to all who stop by — including the groups of law-abiding, recreational motorcyclists (who've certainly conquered the weekend parking problem). A new free trolley also links underutilized "uptown" lots with the historic district. Among its shopping surprises: a **Pindar Vineyards** store where you can taste — and buy — wine seven days a week because the state ban on Sunday alcohol sales exempts wineries and their retail outposts as a sop to small vintners (and tourism).

While shopping and eating are prime tourist goals, the village also offers an eclectic assortment of historical landmarks, from red London phone booths to a thriving theater in a historic vaudeville house (there also is, of course, facing the harbor, the requisite statue of **Thomas Jefferson**).

Most homes of early seafaring families remain private residences, but you can admire the distinctive exteriors (one with a porthole window, another with a porch that lends the aura of a Mississippi riverboat) as you stroll along Main, East Main and their steep side streets (East Broadway was a venue for auto hill-climbing contests in vogue in the early 1900s). Port Jeff's oldest structure, **Roe House**, was built by a shoemaker in 1682 and relocated 300 years later (its second move) to the present site next to the ferry dock, where it serves as the tourist information center. There you can pick up walking tour brochures prepared by the historical society, whose own appropriately vintage headquarters is the former home of leading 19th century shipbuilder **John R. Mather**.

Mather House (built in two stages in the mid-1800s) and its outbuildings are open to the public for tours. The tool shed displays an array of early household and shipbuilding implements — some familiar, some not (you'd never guess the use of the hefty "holy stone"). There's a replica country store, barber shop and butcher shop (five steaks for $1.40); a diorama of Port Jefferson Harbor circa 1900 and a mock-up of **Wilson's Sail Loft** (sail maker for the original America, for which the coveted racing cup was named). Still in the works is a clock museum with about 275 timepieces. A small house next door contains furnishings of a typical pre-Civil War home, as well as the museum's consignment shop.

Port Jeff's grand harbor naturally figures in various annual events. Summer activities include free waterside concerts and fireworks — the latter not scheduled until whenever the local shore birds finish nesting. ◆

Newsday Photo / J. Michael Dombroski

"The Sailor," a link to the village's seafaring past, overlooks the marina that now draws recreational boaters to Port Jefferson.

AT A GLANCE

Chamber of Commerce, 118 W. Broadway, 631-473-1414, www.port jeffchamber.com. **Hours**: 10 a.m. to 4 p.m. weekdays year-round (also noon to 4 p.m. weekends Memorial Day weekend through September); **Mather House Museum**, 115 Prospect St., 631-473-2665, www.portjeffhistorical.org. **Hours**: 1 to 4 p.m. weekends Memorial Day to Labor Day (also Tuesday and Wednesday in July and August). Partial wheelchair accessibility villagewide. Child appropriate.

Did You Know?
Port Jefferson was Suffolk's largest shipbuilding center in the 1800s, when it produced almost half the county's ships.

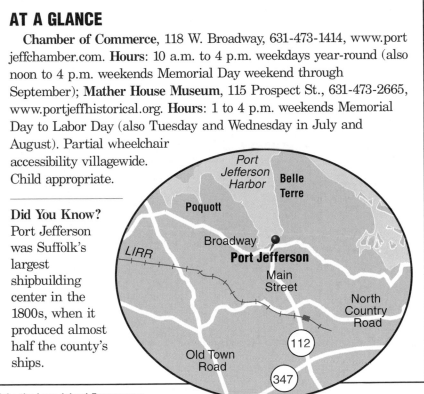

Take the Long Island Expressway to Exit 64 north (Route 112). Continue on Route 112 for about 12 miles to Port Jefferson. From Middle Country Road (Route 25) take Route 112 north approximately seven miles from Coram.

Mount Misery and Other Tales

TIRED OF WALKING? You can take a narrated harbor cruise among the skittering sailboats and listen to some intriguing tales about early Port Jeff. Such as:

Overlooking the village of Drowned Meadow was a high bluff called **Mount Misery** (what was it with these people?). In 1906, the area was re-named Belle Terre by a visionary developer who hoped to create another Newport there. He went bankrupt, but not before he'd persuaded archi-

Newsday Photo / J. Michael Dombroski

The paddleboat Martha Jefferson sails from spring to autumn.

tect **Stanford White** to design a clubhouse (long since burned down), pergolas (swept away in the 1938 hurricane) and a gatehouse — which still guards the exclusive community.

Below the bluff is **Pirate's Cove,** created long after Capt. Kidd's day by dredging companies — though it was a treasure trove, of sorts. During Prohibition, barges pulled in there to stash liquor in loads of sand they were taking to New York City for building projects.

Across the harbor there once was a revolutionary-era British stronghold — remembered by local historians only by the colonists' derisive name: **Fort Nonsense.** (In the War of 1812, Americans manned its cannons.)

Such yarns add an educational component to the numerous daily sightseeing cruises on the 85-foot, 149-passenger paddle wheeler **Martha Jefferson**, which also has summer dinner cruises.

The **"Port Jeff Ferry"** promotes a weekly crossing to Bridgeport in

WHILE YOU'RE THERE

At the nearby State University at Stony Brook, the **Museum of Long Island Natural Sciences** in the Earth and Space Science Building, 631-632-8230. **Hours:** Open 9 a.m. to 5 p.m. weekdays year-round. **Fee:** Small fee for programs only. Exhibits deal with the relationship of mankind and nature on Long Island, especially involving the seashore. Includes dioramas, changing exhibits, nature walks and lectures, some stargazing evenings. The **University Art Gallery** in the Staller Center for the Arts, 631-632-7240. **Hours:** Open Tuesday to Friday noon to 4 p.m., Saturday 6 to 8 p.m., usually year-round, (closed for renovations mid-May through summer 2002); **Stony Brook Union Art Gallery**, 631-632-6820. **Hours:** Open noon to 4 p.m. weekdays during the school year (late-August to mid-May).

Thompson House, 91 North Country Rd., Setauket, 631-692-4664. **Hours:** 1 to 5 p.m. Friday to Sunday, Memorial Day to Columbus Day. **Fee:** $3, $1.50 ages 7 to 14 and over 62. This 1700 saltbox structure has one of the finest collections of early Long Island furniture, dating to 1750, plus a colonial herb garden.

Sherwood-Jayne House, 55 Old Post Rd., East Setauket, 631-692-4664. **Hours:** By appointment late May to early October. **Fee:** $3, $1.50 ages 7 to 14 and over 60. This 18th century colonial farmstead features barns, a corn crib, ice house and sheep.

For a sampling of Port Jefferson area restaurants, see Page 290.

summer as a family sunset cruise (Wednesdays 6 to 9 p.m.), featuring a variety of live musical entertainment. The ferry also has daily casino tours to Foxwoods and Mohegan Sun in Connecticut and same-day tours to other New England attractions. ◆

Paddle-wheeler Martha Jefferson's hour-long harbor sightseeing cruises are $8 for adults, $7 for seniors, $4 for ages 12 and under; summer dinner cruises are $50 to $60; 631-331-3333, www.marthajefferson.com. Bridgeport & Port Jefferson Steamboat Co. Wednesday evening cruise with DJ dance music, $13 adults, $8 ages 6 to 12; daily cruises to a Connecticut casino are $32; 631-473-6282 or www.bpjferry.com.

ROBERT MOSES STATE PARK
AND CAPTREE STATE PARK

Here's The Skinny On Dipping

WHEN THE MAGNIFICENT ocean beach that's now central to **Robert Moses State Park** opened almost a century ago, it was New York State's only public seashore. So the necessary ferry trek was accepted as part of the fun.

Even after the more accessible Jones Beach came along 20 years later in 1929 — with its boardwalk, band shell and sporting activities — escapists still preferred the more remote outer island (some, perhaps, partly for the skinny-dipping, but more about that later).

Some romantics, however, eventually started to like the idea of more convenience.

In 1954, a causeway opened to the halfway point — **Captree Island** — and 10 years later the mainland road link was complete. Sun worshipers for a while had it all: Their beloved barrier beach remained pristine, but they now could zip over to it in a flash (except maybe on Labor Day and extra-hot summer Sundays).

Between the opening of the mainland bridge and the hoopla accompanying the renaming of the beach to **Robert Moses State Park** (in honor of the Long Island State Parks Commission's first president), attendance quickly surged

A perfect August beach day at Robert Moses State Park

to the point that new bathing areas had to be provided.

In 1968, the landmark 202-foot-tall water tower was erected (similar to the Venetian-style campanile at Jones Beach, it stores well water that supplies the park) and other attractions gradually were introduced.

Today the park offers — in addition to five miles of ocean beach for swimming and surfing — an 18-hole pitch-putt golf course and a day-use boat basin that can accommodate 40 boats (fees are charged for both of these amenities), as well as picnic areas, surf and pier fishing, gift shops, refreshment stands, and various cultural and educational activities.

The **Fire Island Lighthouse**, a 0.7-mile walk eastward from the eastern end of Parking Field 5, is the most popular magnet. Another frequent goal of beach walkers — and about the same distance east of the lighthouse — is **Kismet**, the first in the string of Fire Island residential communities that are reachable to the public only by ferry.

Lighthouse Beach is the unofficial name of a clothing-optional stretch of sand to the east and west of the guess-what. Signs announce the probability of encountering nude sunbathers in this popular and well-patrolled section of **Fire Island National Seashore** (in a marked area right in front of the lighthouse, anyone walking is asked to cover up).

While swimming and sunbathing take priority at Robert Moses State Park, across the causeway at **Captree State Park** fishing now is, well, the lure.

During the Gay Nineties, Captree Island was home to New York's famous Tile Club, to which some of America's most celebrated artists and writers belonged. Members' paintings of its windswept beaches hang in many homes and galleries here and abroad.

Captree's fishing fleet — which grew from eight charter and party boats when the park opened to the public in 1954 to several dozen in 2001 — annually carries thousands of anglers into **Great South Bay** or farther into the Atlantic. The park also has a boat launch ramp for private owners.

In addition, Captree's two 10-foot-wide fishing piers accommodate an estimated 250,000 annual anglers — sometimes packed in like sardines in mustard sauce. There are picnic areas nearby so families can pop their catch right on the barbie. The 298-acre park also has refreshment stands and a seasonal full-service restaurant for those whose luck with the rod and reel isn't so good. (Flynn's restaurant in the Fire Island community of **Ocean Bay Park** offers a seasonal package including a surf-and-turf dinner buffet and a boat trip there and back from Captree marina.)

But since you can fish at the park around the clock (night permit required), few persistent anglers go home with only tales of the one that got away. ◆

Newsday Photo / Ghazalle Badiozamani

A brother and sister cool off at the Robert Moses beach.

Newsday Photo / J. Michael Dombroski

A lone seagull perches atop a light pole on a pier at Captree State Park.

AT A GLANCE

Robert Moses State Park and **Captree State Park**, call 631-669-0449 for information on both.

Hours: Open from sunrise to sunset daily year-round.

Fee: Parking fees ($7 Moses, $5 Captree) are collected according to seasonally changing schedules. Partially wheelchair accessible. Child appropriate.

Did You Know?

Until 1920, a telegraph station near Fire Island Lighthouse sighted incoming ships from its watchtower and relayed the news to New York City.

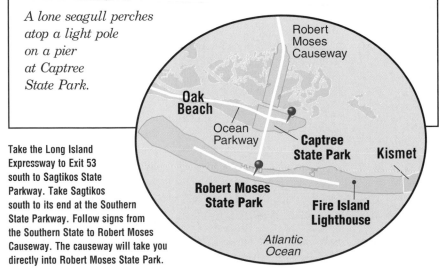

Take the Long Island Expressway to Exit 53 south to Sagtikos State Parkway. Take Sagtikos south to its end at the Southern State Parkway. Follow signs from the Southern State to Robert Moses Causeway. The causeway will take you directly into Robert Moses State Park.

See the Light

FIRE ISLAND LIGHTHOUSE tests visitors' stamina twice.
There's, of course, the daunting 192-step staircase spiraling to the breezy viewing platform near the top. But first there's the walk from the parking lot — 0.7-mile across the dunes.

Both challenges are a snap for anyone who is reasonably fit, however. And they're well worth a little effort. Boardwalks protect the dunes from tramping tourists and are dotted with informative placards about the shore environment and its wildlife. You're almost guaranteed to see at least one deer nibbling brush or even sauntering across your path on your stroll to and from the lighthouse — which rises out of the sand about halfway between Parking Field 5 of Robert Moses State Park (reached by a causeway) and **Kismet** (westernmost of the string of resort communities within Fire Island National Seashore).

The 168-foot white lighthouse tower, marked by two black bands, was built on the brink of Fire Island Inlet in 1858 after the nearby 1826 original was deemed too short, at 80 feet, to adequately warn ships about sandbars. While the lighthouse hasn't moved an inch, the ever-shifting sands had deposited a re-

Photo by John Griffin

The Fire Island Lighthouse, just beyond the east end of Robert Moses State Park

196

markable five miles of land between it and the inlet by the turn of the millennium (Fire Island's westward march continues, at an annual rate the National Park Service likens to 25,000 people daily hauling little red wagon-loads of sand to **Democrat Point**, at the inlet edge).

When you step out onto the lighthouse's railed granite ledge, you can see far beyond Democrat Point — often to New York City's skyline. On your way down through the core of the tower, try to guess why the metal steps are lacey rather than solid (hint: the interior walls were whitewashed for the same reason).

A couple of scenes were shot at the lighthouse for "Men in Black 2," the sequel to Will Smith and Tommy Lee Jones' otherworldly 1997 hit film. So before you descend, keep an eye on the sky. ◆

Fire Island Lighthouse, 631-661-4876. Hours: Open 9:30 a.m. to 5 p.m. weekends and holidays April through June, daily July to Labor Day; 9:30 a.m. to 4 p.m. weekends and holidays Labor Day to mid-December. Fee: Tower tours must be booked in advance and cost $4 adults, $3 older than 65 and younger than 12 (climbers must have appropriate footwear and be at least 42 inches tall). Parking is in Robert Moses State Park Field 5 (fee $7).

Fire Island Lighthouse.
Opposite Islip, Long Island, N. Y.

Nassau County Museum Collection, Long Island Studies Institute

A view of the lighthouse from about 1940

Newsday Photo / Daniel Goodrich

A hiking path at Gardiner County Park in Bay Shore

Newsday Photo / Jim Peppler

Deer hop a fence at the South Shore Nature Center in East Islip.

WHILE YOU'RE THERE

Gardiner County Park, South Country Road, Bay Shore, 631-854-0935 (seasonal) or 631-854-4949. This 231-acre facility offers physical fitness and nature trails.

Islip Greenbelt County Park, main access is through Lakeland County Park or Hidden Pond Park, Islip, 631-854-4949 or Greenbelt Trail Conference, 631-360-0753. **Hours**: Open year-round. **Fee**: Free. This park consists of 153 acres with links to portions of the 34-mile Greenbelt hiking trail.

South Shore Nature Center, off Bayview Avenue, East Islip, 631-224-5436. **Hours**: Open 9 a.m. to 5 p.m. daily through October, then only weekdays until the beginning of April. **Fee**: Free, except for family programs. The center features 206 acres with 2.3 miles of self-guided trails, a nature museum (with saltwater fish tank and reptiles), indoor honeybee hive, butterfly garden, picnic area, weekend family programs (fee).

For more about Fire Island, see Page 66.

For a sampling of nearby restaurants, see Bayport-Patchogue area on Page 280 and Sayville on Page 293.

SAG HARBOR

'Un-Hampton,' But Hardly Unsung

ONE MORE CELEBRITY and Sag Harbor might have to stop calling itself the "un-Hampton."

Writers and other artistic types have been drawn by the workaday simplicity of this old port on the South Fork's north shore since James Fenimore Cooper arrived in 1819 to invest in a whaling ship and scribbled sea yarns in his spare time. Over the decades, the literary set gradually added a patina of chino-chic, as did the stage-and-screen crowd just across North Haven Bridge at the actor's colony (whose present-day residents include singer **Jimmy Buffett**).

Now, anybody you meet on a flight to Paris or Peoria knows Sag Harbor is where **Billy Joel** moors his boat, where playwright **Lanford Wilson** or novelist **E.L. Doctorow** might stroll into the deli while you wait for your coffee and where you can see as many stars in the audience as onstage at the Bay Street Theatre (which is guided by Julie Andrews' daughter, Emma Walton; her husband, Stephen Hamilton; and Richard Burton's ex-wife Sybil Christopher).

Every romantic has found in Sag Harbor a different muse. The legendary choreographer **George Balanchine,** who never lived there, supposedly chose to be buried in the village because it reminded him of the south of France (it's also said that one night a year since he died in 1983, a mystery limo has swept up to his grave in **Oakland Cemetery** to leave flowers).

Sag Harbor's circa-1790 Custom House was the tax collector's home when the village was among America's first offical ports on entry.

The 1846 American Hotel on Main Street dates to the whaling boom.

Yet for all this heady charisma, Sag Harbor somehow manages to keep its focus as a normal place where normal people live and work (albeit sometimes at incredibly interesting things).

The village cinema screens only art films, and there's a modest handful (though not dozens) of antiques shops, galleries and hot restaurants such as the eponymous one run by Oprah-esque dynamo **B. Smith**. But the fire whistle still blows at noon — a throwback to the days when workers didn't wear watches — and Sag Harbor essentially remains a community of small shops with wood floors and reasonable prices (and a minimum of lobster-claw throw pillows). The old-fashioned five-and-dime may not always live up to its billing, but the 7-cent copy machine would be an incomparable bargain if you stopped in to duplicate a village map and found, say, "Hannibal" author **Thomas Harris** running off another sequel to "Silence of the Lambs."

Resident celebrities aside, Sag Harbor has a boom-and-bust history as long as Main Street. Park in one of the free village lots (boaters can tie up at **Long Wharf** for a fee) and pick up brochures at the waterfront windmill or at most any store. Then walk.

The 1844 Old Whaler's First Presbyterian Church on Union Street, viewed from the Old Burying Ground

Sag Harbor's wealth of imposing homes and public buildings stems mainly from its heyday as a whaling capital in the 1800s — a story eloquently told in the **Sag Harbor Whaling and Historical Museum**. Indeed, despite a series of devastating fires, so many noteworthy buildings remain — from early colonists' capes to the Georgian, Italianate, Greek and Egyptian Revival mansions of **"Captain's Row"** — that the whole town is on the National Register of Historic Villages.

You can browse for antiques at the gambrel-roofed **"Umbrella House,"** reputed to be the oldest dwelling; it quartered British soldiers during the Revolution and has cannon shot scars from the War of 1812. And you can attend concerts (or services) at the 1844 **Old Whaler's Church**, the only Sag Harbor structure singled out as a National Historic Landmark; the soaring steeple once visible to ships rounding Montauk Point 25 miles away was toppled by the 1938 hurricane, but one of these days may finally be replaced. When the

AT A GLANCE

The **Sag Harbor Chamber of Commerce** runs an information center at the harborside windmill in the summer, 631-725-0011 or on the Internet at www.sagharborchamber.com. Also helpful: **The Sag Harbor Express** Web site: www.sagharboronline.com.

Did You Know? James Fenimore Cooper wrote about Sag Harbor in "The Sea Lions," Herman Melville did so in "Moby Dick" and John Steinbeck in "The Winter of Our Discontent."

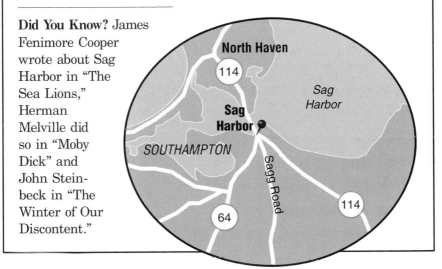

Take the Long Island Expressway east to Exit 70.
Go south on Route 111 to Route 27 (Sunrise Highway). Take Route 27 east to Bridgehampton. Turn left onto Bridgehampton-Sag Harbor Turnpike (Route 64) north. Go about five miles into Sag Harbor.
Coming from Montauk, take Route 114 north at East Hampton. Go about five miles into Sag Harbor.

noon whistle blows (tourists lose track of time, too), consider having lunch — or at least a peek — at the **1846 American Hotel,** the town's classiest lodging.

After whaling declined, Sag Harbor re-invented itself as an industrial center. The ivy-covered brick skeleton of the old Bulova Watch Case Factory, once the region's largest employer, stands as testimony to the passing of yet another era (a condo conversion reportedly is on the horizon, however — in keeping with the village's enduring identity as a resort).

Locals keep a few secrets, such as favorite swimming holes and clamming flats, but readily publicize the world-class sunsets at **Long Beach** and the frequent art openings, pancake breakfasts and authors' readings at comfortably frayed **Canio's** and at newer **Paradise Books** in **The Paradise** cafe. At any such events, half the room is guaranteed to think *you're* a celeb. ◆

Of Whaling and Then Some

STEP BETWEEN the jawbones of a whale at the **Sag Harbor Whaling and Historical Museum** and you enter the world of the 19th century seafarers who put the village on international maps (back when its scruffy waterfront was often considered more dangerous than life at sea).

The imposing Greek temple-front mansion, its roofline ornamentation simulating whaling tools and whale teeth, was built in 1845 for shipowner Benjamin Huntting when his industry seemed to promise everlasting prosperity (sort of like a dot-com). But by the end of the 19th century, it was on the skids — victim of a triple whammy: the cost of chasing fewer whales in distant oceans and the discoveries of petroleum in Pennsylvania (providing a cheaper source of fuel oil) and gold in California (diverting seamen as well as whaling ships, which now carried them west to seek an easier fortune). The mansion then became the summer home of Sag Harbor's greatest benefactor, philanthropist Margaret (Mrs. Russell) Sage — at one point one of America's 10 wealthiest women.

Newsday Photo / J. Michael Dombroski

The Sag Harbor Whaling and Historical Museum on Main Street preserves the salty stories of a seaside village.

Outside are several large kettles called try-pots, used on ships to "try-out" whale oil from blubber, and a replica whaleboat in which six men would battle a whale five times its size. (American Indians, who taught white settlers how to harvest food and oil from whales, simply herded the animals into shallow water to die.)

Not counting changing special exhibits, the museum's conglomeration of artifacts includes ship models, logbooks, harpoons, navigation instruments

Newsday Photo / Bill Davis

At the whaling museum, a half model of a Sperm whale is framed by tools of the trade: harpoons, left; toggle harpoons, right, and blubber spades.

Sag Harbor Whaling and Historical Museum

A painting at the whaling museum depicts "fire in the chimney" — whaler's talk for blood spurting from a wounded whale's blowhole.

WHILE YOU'RE THERE

Custom House Museum on Main Street, 631-692- 4664. **Hours**: May to June and September to October Saturday and Sunday 10 a.m. to 5 p.m.; July and August, Tuesday to Sunday 10 a.m. to 5 p.m. **Fee**: $3 adults; ages 7 to 14 and seniors $1.50. Groups by appointment only. The circa-1790 museum offers a glimpse of how 18th century tax collector Henry Packer Dering, his wife and nine children lived.

The **Sag Harbor Fire Department Museum,** Sage and Church streets, 631-725-0779. **Hours**: Open daily 11 a.m. to 4 p.m.; tours by appointment. **Fee**: $1 adults; 50 cents ages 10 to 14. The department is the oldest volunteer fire department in the state. On display: equipment dating to the 19th century, a 1929 Model T fire chief's truck and a mural depicting some local fires. There are also plenty of kid-friendly bells and whistles.

Temple Adas Israel (1898) at Elizabeth Street and Atlantic Avenue is Long Island's oldest synagogue.

St. David's A.M.E. Zion Church on Eastville Avenue has been an active center for African-American residents almost continuously since it was built circa 1891.

A landmark in Oakland Cemetery, on Jermain Street, is the **Broken Mast Monument** to whalers lost at sea.

Havens Beach is a short walk from Main Street.

Barcelona Neck, 631-444-0273, a few miles east on Route 114, is a 523-acre wildlife sanctuary with four miles of marked trails. **Hours**: Open year-round except during January's deer season. Bird-watchers can enjoy songbirds, shore birds and waterfowl.

and medical kits (stocked with an unsettling array of cure-alls such as strychnine and handcuffs). Among the delicately carved scrimshaw pieces are decorative objects, toys and everyday essentials ranging from canes to corset stays. The museum's hodgepodge also embraces items unrelated to whaling: treasures such as documents signed by George Washington and Abraham Lincoln as well as a mishmash of more dubious value (early household utensils, Indian arrowheads, rock and bird's egg collections). But

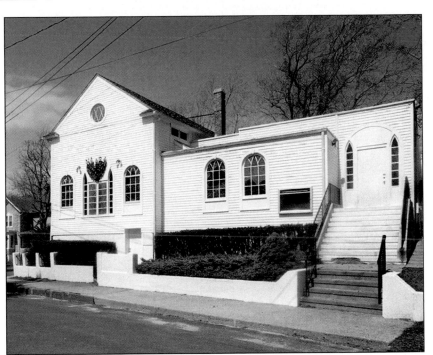

Newsday Photo / J. Michael Dombroski

Temple Adas Israel on Elizabeth Street is Long Island's oldest synagogue in continuous use. It was built in 1898 by European immigrants recruited at Ellis Island to work in a Sag Harbor watchcase factory.

Fee: Access by free seasonal permit, available from the State Department of Environmental Conservation.

For a sampling of Sag Harbor restaurants, see Page 292.

it all adds up to as much fun as a rainy afternoon in Grandma's attic — an aura the museum never wants to change. ◆

Sag Harbor Whaling and Historical Museum, Main and Garden streets, 631-725-0770 or on the Internet at www.sagharborwhalingmuseum.org. Hours: Monday to Saturday 10 a.m. to 5 p.m., Sunday 1 to 5 p.m. mid-May through September. Fee: $3 adults, $2 seniors, $1 ages 6 to 13.

SAGAMORE HILL

Uncle Teddy's Comfortable Home

H E WAS a revered soldier, statesman, sportsman, historian, writer, naturalist and wildly popular president — as well as a sometime trendsetter and catalog shopper. But you really would have loved having Theodore Roosevelt as your Uncle Teddy.

As eloquent a spinner of child riveting dinner table yarns as he was a negotiator of international peace treaties, as enthusiastic a leader of family nature hikes on Long Island as he was a spearhead of African safaris in pursuit of big-game trophies and scientific specimens, Roosevelt the devoted father tackled everyday life as robustly as **Roosevelt the Rough Rider** commanded his volunteer troops in the decisive Spanish-American War Battle of San Juan Hill.

And no place reveals the softer side of T.R. — who would cut short a high-level meeting to keep a promised play date — more than his beloved Oyster Bay residence, **Sagamore Hill**. A tour of the comfortably casual house invariably leaves visitors feeling they, too, could happily call it home (perhaps minus a few dozen animal heads, skins, tusks and other parts).

While Theodore pursued the limelight, his wife, Edith, was content managing Sagamore Hill's farming enterprises — and the brood that soon in-

Theodore Roosevelt's beloved home, Sagamore Hill, built in 1884

cluded Theodore Jr., Kermit, Ethel, Archibald, Quentin and Alice (Roosevelt's daughter with his first wife, Alice, who died shortly after giving birth). Both parents regularly joined the clan, including assorted cousins who added to the general chaos, in everything from joyful beach outings to solemn processions to the backyard pet cemetery (created after Edith noticed guinea pigs being buried in the sandbox). These romps let T.R. be the rough-and-tumble kid that a sickly childhood had prevented. He also pitched in with chores, sometimes disastrously (he once felled a tree that doubled as a telephone pole, knocking out service to the house in his zeal to reclaim the harbor view).

History was made there, of course. It was on Sagamore Hill's broad "piazza" that he was notified of his nomination for governor of New York in 1898, vice president in 1900 and president in 1904 (he eventually removed a section of porch railing so he could give speeches closer to the crowd). The home was his **Summer White House** from 1902 to 1908, and one August day in 1905 Roosevelt brought together the envoys of warring Russia and Japan in his study (and part-time Oval Office) — which earned him the Nobel Peace Prize.

Private and public life merged in the North Room, still bursting with everything from the family piano to gifts from world leaders (including ancient samurai swords from Japan's emperor) and favorite mementos such as Frederic Remington's **"Bronco Buster"** statue, presented to him at Montauk's Camp Wikoff by his disbanded regiment. Roosevelt's Rough Rider hat and saber hang on the antlers of a mounted elk head.

Providing museums with countless exhibit and study specimens enabled Roosevelt to reconcile big game hunting with his conservation bent (he set aside more parkland than any president except Jimmy Carter, who created Alaska's mammoth Denali). Oh — the "teddy bears" in the house? Brooklyn toy makers coined the term after T.R. refused to shoot a cub that hunting guides had tied to a tree so he wouldn't go home trophyless (never mind that he hated the nickname Teddy).

In the third-floor retreat called the **Gun Room**, note the hippopotamus foot inkwell (which looks just like a modern CD player). It must have been unusual even in his time; T.R. ordered it from a taxidermy catalog.

AP Photo

The study at Theodore Roosevelt's Sagamore Hill estate in Oyster Bay

Roosevelt bullied his way through a lifelong string of health challenges from asthma to an assassin's bullet (you probably could add high triglycerides, given his favorite meal of fried chicken with gravy), but he was vulnerable in matters of the heart. When his youngest son, **Quentin**, was killed in action during World War I, a friend observed that the boy in T.R. also died. Six months later — on Jan. 6, 1919, at age 60 — T.R. died peacefully in his sleep at home.

"I wonder if you will ever know how I love Sagamore Hill," he told his wife the night before he died. She surely knew; every visitor does. ◆

AT A GLANCE

Sagamore Hill National Historic Site, Sagamore Hill Road, Oyster Bay, 516-922-4788, www.nps.gov/sahi. **Hours**: Open Wednesday through Sunday; daily in summer. Free entry to grounds; house tour tickets $5 adults, age 16 and younger free; tickets first come, first served. **Orchard House Museum**, Theodore Roosevelt Jr.'s Georgian-style house on the grounds, offers a short film plus Roosevelt family exhibits; the museum is scheduled to close for renovations January through Labor Day, 2002. Partially wheelchair accessible. Child appropriate.

Sagamore Hill National Historic Site Photo

An elephant tusk and a portrait of the Rough Rider

Did You Know?
In January 2001, Theodore Roosevelt was finally awarded a posthumous Medal of Honor for bravery in 1898 during the Spanish-American War.

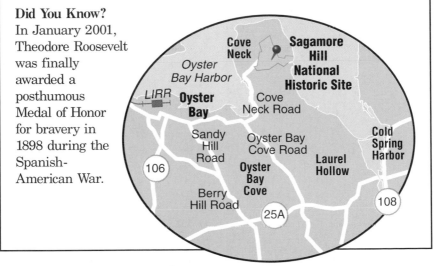

From Long Island Expressway Exit 41N, take Route 106 north eight miles to Cove Neck Road in Oyster Bay Cove. Make a right on Cove Neck Road and go about one mile to Sagamore Hill Road, on right. Sagamore Hill is a short distance ahead.

This Page, Sagamore Hill National Historic Site / National Parks Service Photos

A bear rug in the Gun Room on the third floor at Sagamore Hill. Other curiosities include an African drum and a hippopotamus foot inkwell.

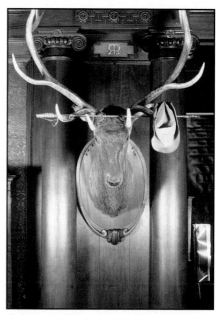

Trophies keep watch at the door to the Sagamore Hill dining room.

A Rough Rider hat and sword decorate a North Room trophy.

T.R.'s Final Resting Place

Y OUNGS CEMETERY, a short walk down the road from Sagamore Hill, was selected by Theodore Roosevelt and his wife as their
final resting place perhaps because of their friendship with an Oyster Bay neighbor.

The cemetery began as the burial ground for the family of one of the area's early 17th century colonists, **Thomas Youngs**. Over the years, nine generations of Youngs lived in
the homestead across the
street — where George
Washington stayed during a
presidential tour of Long Island. The last male in the
Youngs line, attorney **William Jones Youngs**, served as
Roosevelt's secretary during
his term as New York governor.

Newsday Photo / Michael E. Ach

T.R. and Edith's simple
grave site, enclosed by a
high wrought-iron fence, is
at the top of 26 steps
(which number president
was he?). A family section
elsewhere in the cemetery
includes the graves of son
Archie, daughter Ethel and
their spouses. By family tradition, Roosevelts are buried where they fall, so the
rest of the family is scattered. Quentin was buried

*The grave of Theodore Roosevelt and his
wife Edith in Oyster Bay*

where he was shot down in German-occupied France, but in the 1950s the
family asked that he be reinterred in the National American Military Cemetery in Normandy beside his brother Ted, who died of a heart attack a few
weeks after directing the D-Day landing on Utah Beach. Kermit, who com-

Newsday Photo / Nelson Ching

A great horned owl at the Theodore Roosevelt Sanctuary

mitted suicide in Alaska, is buried there. Alice, who kept Washington society on edge with her acerbic observations and was the last child to die (at age 96, in 1980), is buried in the capital. T.R.'s first wife, Alice, is buried in a Roosevelt family plot in Brooklyn's Greenwood Cemetery.

Theodore Roosevelt Sanctuary, adjoining Youngs Cemetery, became the first National Audubon Society songbird sanctuary when it was founded in 1923 on land donated by cousins of T.R. with the condition that it be preserved in his memory. It's also the home of a memorial fountain commissioned by the National Association of Audubon Societies after his death.

The sanctuary offers hiking trails; a small museum with educational programs, events and exhibits, plus an aviary that is home to birds that can't be released due to permanent injury (recent residents have included a bald eagle, red-tailed hawks, great horned owls and eastern screech owls). ◆

Theodore Roosevelt Sanctuary, 134 Cove Rd., 516-922-3200. Hours: Open 8:30 a.m. to 4:30 p.m. weekdays, 9 a.m. to 5 p.m. weekends.

Photo by Thomas A. Ferrara

Raynham Hall features 18th century and Victorian-era furnishings.

WHILE YOU'RE THERE

Raynham Hall, 20 W. Main St., Oyster Bay, 516-922-6808,
www.raynhamhallmuseum.org. **Hours:** Tuesday to Sunday 1 to 5
p.m., summers noon to 5 p.m. **Fee:** Adults $3, seniors / students $2,
6 and younger free. The circa-1740 structure was the home of Quaker
merchant Samuel Townsend, an ardent patriot, whose son was one of
Gen. George Washington's spies during the Revolutionary War (even
while the family was forced to host occupying British troops during
the winter of 1778-1779). Features 18th century and Victorian-era
furniture. The garden is also open to the public.

Oyster Bay Historical Society, 20 Summit St., Oyster Bay,
516-922-5032, www.members.aol.com / obhistory. **Hours**: Research
library and museum 10 a.m. to 2 p.m. Tuesday to Friday, 9 a.m. to 1
p.m. Saturday, 1 to 4 p.m. Sunday. Closed Monday. **Fee:** Donation.
Features the circa-1720 Earle-Wightman House and restored 18th
century garden; exhibits include an opportunity for kids to dress in
revolutionary garb and other hands-on history projects.

Tiffany Creek Preserve, Sandy Hill and Berry Hill roads, Oyster
Bay Cove. Run by Nassau County, 516-571-8500. **Open**: Year-round.
Fee: Free. This 197-acre preserve, composed of parts of three former
estates, was acquired by Nassau County in 1992. Map-brochures to
a self-guided wooded trail are available in the parking lot.

**For a sampling of nearby restaurants, see Cold Spring Harbor on
Page 282 and Oyster Bay on Page 290.**

ST. JAMES

AND DEEPWELLS FARM

Time Traveling To The Early 1900s

NEW YORK CITY'S 94th first lady is having a celebrity friend to tea at her Long Island farmhouse and you're invited, too.

They won't be there in person, of course, because she and her contemporaries are long gone (the city is now up to mayor No. 108). But they're vividly portrayed by local actresses in the ongoing "Biographies on Stage" series presented at **Deepwells Farm** in St. James.

Deepwells was the country home of **William J. Gaynor**, who served as New York's mayor from 1910 until his death in 1913 (of a heart attack believed related to injuries suffered in an assassination attempt during his first year in office). Gaynor loved the circa-1845 farm he bought in 1905, making pets out of the livestock such as a favorite sow named Nancy (he said politicians could learn a lot from pigs). While he was recuperating from his gunshot wound, he ran the city from there.

His spirited wife, **Augusta Gaynor** — who had studied opera and occasionally performed at benefit concerts — also loved the St. James area, which had attracted a colony of silent film stars and other notables.

Sitting at tea tables in the parlor, you learn about their lives and times

A performance in the "Biographies on Stage" series at Deepwells Farm

via witty 90-minute dramas written by Long Islander Sal St. George. The one-act plays focus on **Augusta**, her outspoken head housekeeper, Isabel, and some prominent visitors (primarily women, like the audience). Mrs. Gaynor often sings for her guests (and may ask them to join in). Isabel invariably brings up some favorite treat or new household product she's found at **St. James General Store** just down the street.

Established 48 years before the Gaynors bought their homestead, this intriguing emporium is billed as the oldest continuously operated general store in America. Like Deepwells, it's now owned by Suffolk County, operated by **Friends for Long Island's Heritage** and listed on the National Register of Historic Places.

In the early days, it sold everything from clothes and groceries to horse medicine and hardware, and served as the local post office and social center, to boot. Store ledgers show signatures of a staggering array of visitors from **Irving Berlin** to heavyweight champion **James J. "Gentleman Jim" Corbett**. As in those days, craftspeople still drop by to share skills such as quilting, chair caning or jelly making on a weekend afternoon. And musicians again play on the front porch during summer craft fairs.

While some of the wares that cram its two floors are the familiar stuffed animals and bubble bath you find at gift shops of decidedly more

Newsday Photo / Tony Jerome

The St. James General Store — in continuous use since 1857

recent vintage, the hodgepodge is worth careful scrutiny. Quality home and garden items, plus a large stock of books on Long Island history, share space with souvenir mugs, T-shirts and Victorian-era high-button shoes and bonnets. Prices may keep up with inflation (penny candy starts at a dime), but in line with the aim to reflect the period from 1880 to 1910, saleswomen dress as **Gibson Girls** — then considered the epitome of feminine fashion.

The store is unchanged structurally since 1894, and the counters, display cases, coffee grinder, potbelly stove, produce barrels and checkerboard are originals. Almost-original furnishings include the post office window (all postal transactions are still authorized there but time constraints limit the store to stamp sales) and, near the portrait of founder **Ebenezer Smith** on the front wall, an antique crank phone. It may or may not occupy the exact spot as the one supposedly fried by a lightning bolt moments after another famous St. James habitue — architect **Stanford White** — had finished a call one stormy day.

There's no one left to verify the truth to such tales. But whether fact or fiction, they're as much fun to ponder as the contents of the groaning shelves. ◆

AT A GLANCE

Deepwells Farm, Route 25A (North Country Road), St. James; call 631-862-6080 to book house tours (given noon to 3 p.m. most Sundays for free, though donations are appreciated) and "Biographies on Stage" ($30 per person includes performance and tradi-

Newsday Photo / Ken Sawchuk

tional afternoon tea; "producer's performances" and holiday shows are $35). Partially wheelchair accessible. Partially child appropriate.

St. James General Store, Moriches and Harbor Hills roads (one block north of Route 25A and Deepwells Farm), 631-862-8333.

Hours: 10 a.m. to 5 p.m. daily March through December except Easter, Thanksgiving and Christmas; also closed New Year's Day and Mondays in January and February.

Child appropriate.

Did You Know?
Gaynor named his farm Deepwells after drilling two you-know-whats; it was originally built for a descendant of Smithtown founder Richard (Bull) Smith.

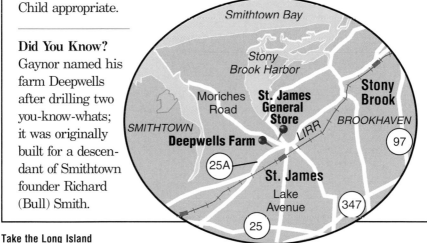

Take the Long Island Expressway to Exit 56. Go north on Route 111 about four miles to Village of the Branch. Take Route 25A east about three miles. Deepwells is on the left. At corner, turn left onto Moriches Road. The *St. James General Store* is a short distance on the right. From the east, take LIE to Exit 62 north. Take Nicolls Road (Route 97) north to Route 25A. Go west approximately five miles to Moriches Road.

Country Elegance

BEFORE THERE was a Deepwells Farm, a St. James General Store — or even a St. James — there was **Mills Pond House**, built a few miles east of these other sites in the Village of Head of the Harbor, which in the early 19th century encompassed a wide area.

The house was designed in 1838 by noted New York architect **Calvin Pollard** for **William Wickham Mills**, who came from a long line of wealthy Long Island farmers. Its big city architect, high-style Greek Revival design (including front and rear stoops supported by Doric columns) and use of imported materials such as English glass and Santo Domingan mahogany, set it apart from other homes in the area then — and now.

Newsday Photos / Tony Jerome

The Mills Pond House in Head of the Harbor, built in a Greek Revival style in 1838, was put on the National Register of Historic Places in 1973.

After descending in the Mills family for generations, it was given to the Town of Smithtown in 1976 and today is home to the Smithtown Township Arts Council. In its contemporary life as **Mills Pond House Gallery**, it's filled year-round with poetry readings, antiques shows, concerts, kids theater, art classes, and exhibits for children and adults. As you browse, you can still appreciate some of the building's original Grecian interior details, such as gold-veined marble mantelpieces and ornamental plaster ceilings — as artistic as the gallery's ever-changing displays. ◆

Mills Pond House, 660 Rte. 25A, St. James; 631-862-6575. Hours: Tuesday to Friday 11 a.m. to 4 p.m., Saturday and Sunday noon to 4 p.m. Fee: Free admission to gallery and outdoor concerts; indoor special events have various fees.

WHILE YOU'RE THERE

St. James is believed to be the only hamlet in New York State that was named after its prominent place of worship — **St. James Episcopal Church**. The 1854 Gothic Revival building has other claims to fame, too. One of its memorial stained glass windows (the Good Shepherd) is a signed Tiffany piece and three others are believed to have been designed by parishioner Stanford White. The noted architect — whose wife, Bessie, was one of the locally prominent Smiths — is buried in the churchyard (the White family plot is in a grove of evergreens, his grave marked by a tall column topped by a carved seashell). The church is at 490 North Country Rd. (Route 25A); 631-584-5560.

A quick detour south of 25A on Lake Avenue takes you to the restored original 1873 Victorian **St. James station house**, the Long Island Rail Road's oldest.

A post-1865 window at St. James Episcopal Church, dedicated to the first senior warden, William W. Mills, and his wife, Eliza A. Mills.

Sweetbriar Nature Center, 62 Eckernkamp Dr., Smithtown, 631-979-6344. **Hours**: Monday to Friday 9 a.m. to 4 p.m., weekends noon to 4 p.m. **Fee**: Donations accepted. The center, on 54 wooded acres along the Nissequogue River, has gardens, trails, a butterfly house (seasonal), guided walks, exhibits and classes. Part of the Greenbelt Trail runs through the preserve.

For a sampling of St. James restaurants, see Page 293.

SANDS POINT PRESERVE

A Page Out Of Fitzgerald

WHERE in the *world*, never mind on Long Island, could you visit an English manor house, an Irish castle and a French country mansion all in one afternoon?

It's possible in **Sands Point** — the likely old-monied East Egg in "The Great Gatsby," F. Scott Fitzgerald's legendary novel of 1920s Gold Coast decadence.

The three buildings are now part of 216-acre Sands Point Preserve — "where history meets nature" via monthly seasonal guided walks of the grounds, tours of sprawling **Hempstead House** (the original estate's Tudor-style main residence overlooking Hempstead Harbor) and hilltop **Falaise** (built in the style of a 13th century Normandy manor, its name the French word for cliff) and exhibits in turreted **Castlegould** (a stable and carriage house modeled on Ireland's Kilkenny Castle and now home to a visitor center, museum shop and changing nature exhibits).

Castlegould — the name initially applied to the entire estate — was created by railroad tycoon Jay Gould's son Howard, who bought the land in 1900 and 1901 and spent more than $1 million developing it. When he moved to England in 1917, he sold the turnkey property to philanthropist **Daniel Guggenheim** for $600,000. In 1923, Guggenheim gave a 90-acre chunk as a wedding present to his son Harry, who built Falaise.

Castlegould, one of two castles built by Howard Gould at Sands Point

Forty-room Hempstead House — with its sunken palm court, gold leaf ceilings and details copied from old palaces — was considered one of Long Island's most opulent **Gold Coast** homes. Eleven rooms on the second floor permanently display a collection of Wedgwood dating from the company's founding in 1759. Otherwise it's been empty since Daniel Guggenheim died in 1930, when his widow moved to a smaller residence on the estate. At the start of World War II, she donated Hempstead House, Castlegould and 162

Falaise was built in the style of a 13th century Normandy manor.

acres to the **Institute of Aeronautical Science**. It was next sold to the U.S. Navy, which set up laboratories there to test electronic devices for submarines, jet aircraft and space flight. Then in 1971, a few years after the testing center moved to Florida, the government deeded 127 acres to Nassau County for recreational use.

That same year Harry Guggenheim died, and Falaise similarly went to the county (which promptly OK'd the filming of two scenes for **"The Godfather"**: a dining scene, shot in the living room, and the infamous horse head scene, shot in the dining room).

The dining table is now set with the official china from Harry's stint as ambassador to Cuba, but the house is otherwise exactly as he left it — filled with centuries-old antiques, important works of modern art and trophies (his racing stable included Kentucky Derby winner Dark Star). A man of many interests, he and his third wife, **Alicia Patterson**, also co-founded Newsday. But his lifelong passion was aviation. He was a Navy pilot in both world wars (because of his age he had to pull strings to see action in the second) and relished being addressed as Captain. He persuaded his father to set up foundations to promote aeronautics and aerospace science, and supported such trailblazers as Orville Wright and rocket pioneer Robert Goddard.

A four decade friendship with **Charles A. Lindbergh** began when Harry

AT A GLANCE

Sands Point Preserve, 95 Middle Neck Rd., 516-571-7900. The preserve's six marked nature trails are open daily 10 a.m. to 5 p.m.; there is a $2 fee for weekend visitors older than age 12. **Castlegould** is open Tuesday to Sunday 10 a.m. to 5 p.m. when there's an exhibit. **Hempstead House** and **Falaise** are open early May to late October (Hempstead House weekends noon to 4 p.m., Falaise Wednesday to Sunday noon to 3 p.m.). Partially wheelchair accessible. Child appropriate.

Did You Know? In addition to "The Godfather" (1972), films shot at Sands Point Preserve include "Great Expectations" (1996) and "Scent of a Woman" (1992).

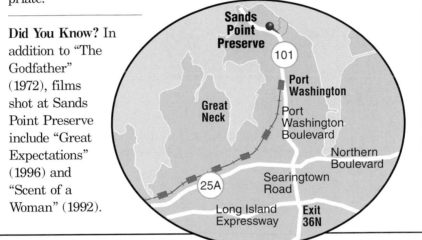

Take the Long Island Expressway to Exit 36N to Searingtown Road. Go north on Searingtown Road past Northern Boulevard (Route 25A); road becomes Port Washington Boulevard (Route 101). Continue on Route 101 for about five miles. The Sands Point Preserve is on the right.

offered him refuge from the public at Falaise after his history-making Paris flight. Lindy's Ford Falcon station wagon still sits in the carport next to Harry's Cadillac, and many of Lindbergh's books fill the shelves — which he hated to see empty after Harry donated his own collections to the U.S. Naval Academy. It was also at Falaise that Lindbergh had one of his first dates with his future wife, **Anne Morrow**, daughter of a business partner of Harry's.

Other than an air show (banned by the deed), what more appropriate event could honor these Renaissance men than the preserve's annual Medieval Festival — complete with a jousting tournament and castle siege? ◆

Newsday Photos / Ken Spencer

The beehive oven in the colonial-era kitchen at the Sands-Willets House

Through the Sands of Time

YOU CAN easily guess the name of one of the first families to settle what's now Sands Point, but you'll learn a lot about the area on a visit to one of the early Sands homes — dating to between 1715 and 1735.

The **Sands-Willets House** (the Willets family moved in about a century later) displays an absorbing collection of artifacts — from an early copy of the 1709 "Gate Rights Map," showing property boundaries, to possessions of "March King" **John Philip Sousa**, a Sands Point celebrity resident during the 1920s. The Sands family owned the imposing white house on Port Washington Boulevard until about 1845 (best known occupant: Revolutionary War hero **Col. John Sands IV**, who led the Great Neck-Cow Neck-Hempstead Harbor Militia).

During an 80-year tenure, Quaker businessman **Edmund Willets** and his descendants left their mark, including the wraparound porch. The property (farmed until 1920) was sold in 1967 to the Cow Neck Peninsula Historical Society — whose name harks back to the days when the hilly landspit now known as the Manhasset Peninsula was relegated to grazing for animals that farmers on the Hempstead Plains didn't want nibbling their crops.

WHILE YOU'RE THERE

Hempstead Harbor Beach Park, West Shore Road, Port Washington, 516-571-7930. **Fee**: Summer parking fee $4 Nassau County residents, $15 nonresidents. This 60-acre seasonal park offers a half-mile of sunny beachfront plus a wooded hillside for picnicking. There's a games area, fishing pier, aerodrome for radio-controlled model airplanes and a paved promenade.

Fishing at Hempstead Harbor Beach Park

Polish American Museum, 16 Belleview Ave., Port Washington, 516-883-6542. **Hours**: Open year-round 9 a.m. to 4 p.m. Tuesday to Friday. **Fee**: Donation. Documents, artifacts and exhibits on achievements of Polish immigrants; gift shop with items made in Poland.

For a sampling of local restaurants, see Great Neck on Page 285 and Port Washington on Page 291.

Another historical society property — and like Sands-Willets, one where children on school trips learn about the good / bad old days — is the circa-1721 **Dodge House**. While the Sands-Willets House has few original furnishings, the seven generations of Dodges who lived in the weathered Dutch colonial at the head of Port Washington's Mill Pond until 1991 seemed to have saved everything — from the tools Thomas Dodge used to build it (the blade of an adze has been matched to marks on the hand-hewn beams) to boxes of kids' homework and 18th century "Old Farmer's Almanacs" to a kitchen restored to 1909, when the room was added.

The three-hole privy out back was filmed by the BBC for a documentary on **Typhoid Mary**, the Irish immigrant who spread the disease to numerous Long Island estates where she worked in the early 1900s. The nearby vegetable garden thrives in its original spot, while a flower garden on the other side of the house is believed to be the final resting place of Thomas Dodge II — who had wanted to be buried outside the front door. ◆

Sands-Willets House, 336 Port Washington Blvd. (corner of Homewood Place), and the Thomas Dodge House, 58 Harbor Rd.; for tour information, call 516-365-9074.

SEA CLIFF AND ROSLYN

Where Grandeur Blends With the Arts

SEA CLIFF'S "painted ladies" might not turn heads every-where. But amid Nassau County's more conventional bedroom suburbs, these colorful Victorian houses are showstopping Follies stars.

A century ago, the artsy one-square-mile community was a booming summer resort. Paddlewheelers regularly steamed in, and a cable car carried visitors up to hilltop lodgings for 10,000 guests — twice the current year-round population. Today the village rests in the past above a tree-shaded shore boulevard a few miles north of busier **Roslyn** — whose own intriguing historical roots reach back to colonial times.

Sea Cliff now attracts lots of architecture buffs with strong legs, who stride up and down the steep streets and public stairways admiring the gingerbread porches and Gothic gables of its landmark **Vickies**. About two dozen village homes are on the National Register of Historic Places and about 900 structures predate World War II — the most of any Long Island community. (None are open to the public except on house tours held every other year, alternating with garden tours.)

A few businesses in the small commercial district along Sea Cliff Avenue also date to its heyday. **Costello's Pub** has been there since 1889, **Arata's** deli since 1903. And locals think jammed-to-the-rafters **Sea Cliff Books**

The setting sun provides a show at Memorial Park in Sea Cliff.

The Woodshed, on Central Avenue between 16th and 17th avenues, is among the Victorian homes that draw tourists to Sea Cliff.

has been there "forever." (You may truly believe you're in a time warp if you spot the resident Theodore Roosevelt look-alike strolling around town.)

Modern-day Sea Cliff began in 1871 as a Methodist summer campground for the new Victorian middle-class. By 1883, when the village incorporated, tents had become cottages and word was spreading about the pretty little resort an hour from New York. By the 1920s, however, autos brought increased freedom to a new generation whose vacation tastes ran more along the lines of Club Med than Club Methodist. So artists bought the drafty old guest houses being scorned as white elephants (there's still an annual October arts and crafts fair, and December studio-gallery tours.)

This heady past is chronicled in the small **Sea Cliff Village Museum** in the 1920s Gothic Village Hall complex (originally Sea Cliff Methodist Church). The museum has walking-tour maps that pinpoint landmark Vickies such as **The Woodshed**, built in 1890 by a family named Wood (and nearly a century later a setting for a TV movie starring Bette Davis). The village also has two distinctive Russian churches built by emigres who came here during the 1930s and 1950s.

Most spectacular of Sea Cliff's many green spaces is **Memorial Park,** or "Sunset Park," at Prospect and Sea Cliff avenues. Where a grand hotel stood, you can enjoy grand views of Hempstead Harbor and Long Island Sound (in summer supplemented by free Friday twilight concerts) — a reminder that some things never change.

Follow Shore Road south past **Tappan Beach**, zig and zag a bit and you come to Roslyn — where one block of the Main Street historic district boasts 37 structures built from 1690 to 1865, according to a sign near the landmark Clock Tower. Circa-1680 **Van Nostrand-Starkins House**, Roslyn's oldest, is open season-

Newsday Photo / Bill Davis

Roslyn's clock tower, built in 1895

AT A GLANCE

Sea Cliff Village Hall, 516-671-0080, www.seacliff.org. Sea Cliff Village Museum, 95 10th Ave., 516-671-0090. Hours: 2 to 5 p.m. Saturday and Sunday (closed August and September). Fee: $1 requested donation. Roslyn information: 516-625- 4363, www. historic roslyn.org.

Did You Know? Sea Cliff was once part of a vast North Shore farming community, with Roslyn a commercial hub for the rural area.

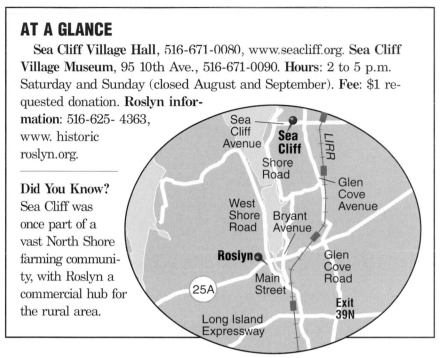

Take the Long Island Expressway to Exit 39N to Glen Cove Road north. Follow Glen Cove Road into Greenvale, just north of Northern Boulevard (Route 25A). Turn left onto Glen Cove Avenue and follow it north for about two miles. Turn left onto Sea Cliff Avenue to downtown Sea Cliff. *For Roslyn* go south from Sea Cliff on Shore Road or take Glen Cove Road to Northern Boulevard (Route 25A), turn right (west) onto 25A and go about a mile to Main Street, Roslyn.

ally and you can peek inside some private homes on annual house tours.

Gerry Park, named for a local preservationist, has a replica of colonial bigwig Hendrick Onderdonk's paper mill — which perhaps provided memo pads to George Washington's spies. On his post-Revolution thank you tour of Long Island, the new president had breakfast at Onderdonk's home (now George Washington Manor restaurant). A nearby Onderdonk grist mill was a tony tearoom in the early 1900s, when culture vultures flocked to local author **Christopher Morley's** summer theater.

A century earlier, when the village was still referred to (with increasing confusion) as "Hempstead Harbor," resident poet-editor William Cullen Bryant is credited with naming it after a favorite spot in Scotland. But he's buried in Long Island's Roslyn, as is Morley. ◆

Picturesque Shore Road and Hempstead Harbor in Sea Cliff

A view of Main Street in Roslyn at dusk

The Nassau County Museum of Art in Roslyn Harbor

A Mansion Filled With Art

T HE WORDS ART AND FRICK are famously linked at the Manhattan mansion-museum bearing the family name. But a Long Island Gold Coast estate that was a Frick home for almost 50 years also is a noted art forum.

Steel baron Henry Clay Frick bought the 145-acre Roslyn Harbor property that's now the **Nassau County Museum of Art** and sculpture-filled **William Cullen Bryant Preserve** for his son and daughter-in-law when they were expecting their third child and still sharing his Fifth Avenue home (he offered jewelry, she said she'd prefer a house, thank you very much).

The Georgian mansion was built about 1900 on what was originally poet and editor William Cullen Bryant's "upland farm" (a gingerbread-Gothic half-ruin of a guest house still standing was built by Bryant, whose own home, **Cedarmere**, can be visited nearby).

When Childs and Frances Frick assumed ownership of the estate in 1919, they redesigned the house and named it **Clayton** — same as the elder Frick's Pittsburgh home. To suit Childs Frick's needs as a paleontologist, naturalist and sportsman, they added a swimming pool, practice polo field, ski slope, laboratory, aviary and bear pit (whose tame occupant often lumbered across Northern Boulevard following the scent of honey to a local beekeeper's). The formal garden Frances Frick helped create — now a sculpture venue like the rest of the grounds — was her grave site until Childs died and the Pittsburgh family cemetery became their resting place.

The museum's two floors of galleries feature changing exhibits plus a permanent collection of 19th and 20th century European and American art. There's a cafe, gift shop and bookstore — plus activities including art classes and trips, lectures and family festivals.

Also on the grounds is the **Tee Ridder Miniatures Museum**, featuring 26 tiny rooms created by Madeline "Tee" Ridder — plus special exhibits, workshops and a gift shop that stocks handcrafted mini-furnishings. ◆

Nassau County Museum of Art, 1 Museum Dr. (off Northern Boulevard), Roslyn Harbor, 516-484-9337 (tape), 516-484-9338 (live), 516-484-7841 (miniatures), www.nassaumuseum.com. Hours: Tuesday to Sunday 11 a.m. to 5 p.m. (closed for two weeks whenever exhibits change). Fee: $6 adults, $5 seniors, $4 students ages 5 and older (rates include admission to both museums). Free docent-led tours at main museum Tuesday to Saturday at 2 p.m.

Newsday Photo / Ken Spencer

The grounds of the Nassau County Museum of Art

WHILE YOU'RE THERE

Cedarmere, Bryant Avenue, Roslyn Harbor, 516-571-8130. **Hours**: Open 1 to 5 p.m. weekends late April through early November. The Long Island home of 19th century poet and editor William Cullen Bryant, who also owned the land that later became the namesake preserve with the art museum, features exhibits, furnished rooms, restored gardens, pond and mill.

Christopher Morley Park, Searingtown Road, Roslyn, 516-571-8113. Walkways and nature trails meander through 30 acres of woods in this 98-acre Nassau County park (for residents and their guests) named for the noted author. In addition to **The Knothole**, as Morley called the cabin where he did most of his writing, there's a jogging course, model-boat basin, playground, picnic areas, playing fields and courts, plus fee-based venues such as a nine-hole golf course, outdoor swimming pools and seasonal ice skating rink.

For a sampling of nearby restaurants, see Sea Cliff-Roslyn on Page 294.

Newsday Photo / Jim Peppler

Skaters circle the rink at Christopher Morley Park.

SOUND AVENUE

Getting On The Route Of the Heartland

FALL'S BRILLIANT REDS and oranges don't just adorn distant mountainsides — they also signal peak picking in East End apple orchards and pumpkin patches.

Along with the chance to pluck your own fruit from trees or vines (and devour homemade jams, pies and such seasonal delicacies as cider and roasted corn), a day in farm country offers hayrides, "maize mazes," wine festivals and other once-a-year activities.

A popular route into Long Island's down-to-earth heartland is **Sound Avenue**, which weaves back and forth through time as it meanders for about 20 miles between its junctions with Route 25A in **Wading River** and Route 25 in **Mattituck.** For most of its length, this gently roller-coastering two-lane byway is a mouthwatering compote of generations-old farm stands and front-yard arrays of homegrown produce, multicolored flowers and other products. Need a new flagpole? There's a hard-to-miss seller along Sound Avenue. Cut your own Christmas trees? They're there, too. A quick side-road detour can hold offbeat attractions such as a pasture where buffalo roam.

Sound Avenue naturally changes with the seasons, and while spring and summer also offer bountiful fresh fruits, vegetables and flowers, a special warmth and excitement surrounds autumn harvest time.

If you start your journey at the western end of Sound Avenue (which

Apple season is a busy season at Briermere Farms in Riverhead.

parallels Long Island Sound, though never close enough for views), the first shopping stops are huge Lewin Farms and Fritz Lewin Farm. Both offer U-pick pumpkins and apples. The next few miles offer The Family Loft gift shop, Rottkamp's Fox Hollow Farm and the small Baiting Hollow Commons shopping complex — which includes a branch of the Hamptons landmark Lobster Roll Restaurant.

You'll spot Briermere Farms by the clutch of cars crowding the small sales store famous for its homemade fruit pies. A board outside the door lists the day's offerings — often a couple of dozen, from ever-popular apple and pumpkin to combos like strawberry-rhubarb.

Then come Gabrielsen's / Helen's Country Plant Farm, Reeve Farm, Sunburst Acres (whose small cafe offers snacks including roasted corn and homemade pie) and Tom Healy's Fruitwood Smoked Fish & Fowl, selling such delicacies as Long Island duck, eels and bluefish.

For a short detour to the Wild West, turn south on Roanoke Avenue and scan the fields around Ed Tuccio and Dee Muma's North Quarter Farm for the herd of American buffalo, or bison. The pastures, also home to several dozen horses, are bounded by Middle Road and Reeves Avenue. Call ahead (631-727-2475) and you may be able to stop by for a tour of the spread, where Dee teaches dressage to numerous celebrities.

Newsday Photo / Tony Jerome

Customers enjoy autumn treats at Harbes Family Farm in Mattituck.

Ready for a sip of Long Island wine (or grape juice)? Palmer Vineyards, one of Long Island's largest, offers an absorbing self-guided tour en route to its pub-like tasting room. Between Palmer and the next winery — Martha Clara — are Northville Farm (touting award-winning cheesecake), North Fork Preserve (a forest of U-cut Christmas trees, waiting for *their* season) and Landscape Adventure (a nursery whose gift shop sells great scarecrows — which humans, at least, will find much more charming than frightening).

Across the road from Martha Clara Vineyards is Sound Avenue's major historic site: **Hallockville Museum Farm & Folklife Center,** open for tours and special events. Next is another country icon: Harbes Family Farm, revered for its rotisserie-roasted sweet corn. There's always something going on at Harbes — also known for its world-class corn mazes here and at its location on Route 25 in Jamesport. Jens Flower Shop across Sound Avenue has great garden gifts and a cluttered sale room where everything is half-price.

The trip ends with Sound Avenue's other two vintners, Macari and Lieb Cellars. Then you can about-face — and hit all the places you didn't notice on the drive east. ◆

AT A GLANCE

Free vacation guides that include Sound Avenue sights are available from the **North Fork Promotion Council**, 631-298-5757 or on the Internet at www.northfork.org. Another handy guide called "Riverhead Any Season . . . For Any Reason" is available free from Riverhead Town Hall, 631-727-3200. Most Sound Avenue attractions are partially wheelchair accessible and child appropriate.

Did You Know?
Suffolk County ranks first in the state for market value of agricultural products sold, though only 49th in acres of farmland.

Take the Long Island Expressway to
Exit 68N to William Floyd Parkway. Go north about
seven miles to Route 25A. Go right (east) on 25A for about three miles to the beginning of Sound
Avenue. This road continues east for about 17 miles before becoming Main Road in Mattituck.

◆

Hallockville's Farm Center

FARMING IS no slacker job, and it was even more arduous back when New York State Militia Capt. **Zachariah Hallock** put down roots at a Riverhead farmstead after the Revolution (perhaps one reason the work is *described* more than practiced there today).

But modern-day visitors to the property — now an 8-acre National Register historic site called the **Hallockville Museum Farm & Folklife Center** — get a clear view back into the two centuries when Zachariah and his family farmed a 102-acre spread. The far-flung Hallock clan is still involved, and hundreds of family members often descend for an annual reunion picnic.

Hallock's house, built in 1765 by an earlier farmer named **Reuben Brown**, kept evolving until 1907 when a porch was added and a wing removed. Furnishings and household items span the continuum — the 1860 kitchen boasts such inventions as a colonial slow-cooker (kept simmering on a bed of hot stones inside a lidded chest). Zachariah's 1820 cobbler shop, once a separate building, is believed to be the state's oldest.

The large English-style barn contains a few old sleighs and carriages but is more importantly home to Emma, a benign Jersey cow, and a heifer, named Hallie. A chicken coop houses a small flock of **Rhode Island Reds**, common to early New England and Long Island. Among the more unusual outbuildings is a four-seat privy, even more noteworthy for its two doors (perhaps to ensure that no one would ever be shut out, or in, by snowdrifts).

In 1900, farmers in the Hallockville-Northville area kicked in to build a pier out into the Sound, hoping ships could get their crops to New York markets more cheaply than by train. But the **Iron Pier** lasted only a few years, succumbing to shifting ice floes during the winter of 1904. Pier Avenue is its only legacy.

The dusty old farm is frequently enlivened by special events that include local "tradition-bearers" such as farmers, baymen and folk artists. Victorian Christmas is celebrated the first weekend of December, a Civil War Camp is held the third weekend in May, a Fall Festival the first Sunday in October. ◆

Hallockville Museum Farm, 6038 Sound Ave., in Riverhead, 631-298-5292. Hours: Open 11 a.m. to 4 p.m. Wednesday through Saturday year-round. Fee: Admission is $5 adults, $4 ages 6 to 16 and over age 65.

Newsday Photo / Tony Jerome

Children dressed for a Hispanic Heritage Festival at the Hallockville Museum Farm & Folklife Center.

WHILE YOU'RE THERE

Wildwood State Park off Sound Avenue (Route 25A) in Wading River (north on Hulse Landing Road), 631-929-4314. **Hours**: Open daily year-round sunrise to sunset. **Fee**: Entrance $7 per car daily in summer, $5 weekends late spring and early fall. This 769-acre site offers a beach, picnic area, fishing, hiking (about 11.5 miles of trails), camping, softball field, basketball courts, horseshoe pits, cross-country skiing and sledding (winter). Wildwood is the only state park on Long Island offering full hookups for trailer campers.

Indian Island County Park, off Cross-River Drive (Route 105) in Riverhead, 631-852-3232. **Fee**: $2 parking for residents, $5 nonresidents Friday to Sunday; various fees for camping and golf. There are 287 acres with hiking, saltwater fishing, picnicking, playground and camping.

For a sampling of nearby restaurants, see Riverhead on Page 292.

Newsday Photos / Ken Spencer

Left, Wildwood State Park in Wading River; right, a totem pole at Indian Island County Park in Riverhead

SOUTHAMPTON

Trendy, Historic And Fruitful

PARIS HAS DEAUVILLE, Rio de Janeiro has Copacabana and New York has **Southampton** — Long Island's irresistible world-class resort, where pool boys have been known to pass for playboys and errant air-kisses easily can be mistaken for a gentle ocean breeze.

Colonized in 1640 by a handful of Massachusetts Pilgrims who landed at **Conscience Point** (long before it became party central), Southampton was the first English settlement on the South Fork. The settlers bought a chunk of bargain-priced land from the native **Shinnecock Indians** (whose intent had been more along the lines of a short-term lease). But all lived a contented agrarian life for the next 200 years. Then rail service arrived in the mid-1800s, bringing the first wave of artists, writers, socialites and gazillionaires. The Indians soon found themselves dispatched to a reservation, and it was full speed ahead for the colonists. Decades bumped along — the Roaring Twenties roaring, the Depression depressing — until post-World War II motels changed the elite summer colony's character once more.

In recent years, the East End potato fields have sprouted vineyards as well as tract homes, and today's South Fork visitors can sample local vintages at three Southampton Town wineries: **Duck Walk** (with several tours daily), **Wolffer** (also home to a cheese company and a stable) and **Channing Daugh-**

242

The shopping district along Jobs Lane in the center of Southampton

ters (whose owner's ingenious wood sculptures adorn the tasting room).

An affluent-cum-egalitarian aura prevails in Southampton's leafy shopping district along Main Street and Jobs Lane, with high-fashion (and -priced) boutiques, noted galleries — plus sprawling **Hildreth's** department store (claiming its 1842 origins make it America's oldest). Between the village and the seven miles of ocean beach is one of the privacy-hedge capitals of the world: the Agawam Lake-Gin Lane estate area. Whereas village denizens such as author Tom Wolfe often are seen strolling about town, comings and goings in these parts tend to be via limo or imported sports car. Still, you never know whom you might spot — say, at seasonal services at nearby **St. Andrew's Dune Church**, its interior adorned with biblical passages, some artfully inscribed by Tiffany (the 1851 building was erected by the U.S. government as a life-saving station, and shipwreck relics dot the grounds).

You'll also find early landmarks in and around the business district (a few open only "by appointment or chance"). The **Southampton Historical Museum** displays antique toys, photos and other items in an 1843 whaling captain's home; the backyard has an 18th century barn and re-created post-Civil War village. The 1648 **Halsey Homestead** is the town's oldest house (also the state's oldest, in Southampton's view, but not all historians agree). The 1750 **Elias Pelletreau Silversmith Shop** next door to the cham-

243

ber of commerce spotlights a colonial patriot's revered craft.

The Italianate-style **Parrish Art Museum**, founded in 1898, offers changing exhibits focusing on American and regional art. Its renowned permanent collection includes major works by **Fairfield Porter** (whose studio was behind his village home) and **William Merritt Chase** (who founded The Art School in Shinnecock Hills, which pioneered teaching the impressionist approach of working totally outdoors rather than sketching outside and completing paintings in a studio). Only the original houses from the art colony that grew up around the school remain, along private dirt roads. But throughout Southampton Town, the creative spirit thrives. Artist **Larry Rivers** has a studio in the village, and Roy Lichtenstein's widow still lives in their house there. Neighboring Bridgehampton and Water Mill are well-stocked with art world luminaries, too.

Cultural hubs? The Parrish is a major venue, along with Roger Memorial Library, Southampton Cultural & Civic Center and Southampton College of Long Island University. Some habitues also would count the many bistros and boîtes. ◆

Newsday Photos / Michael E. Ach

The blacksmith shop at 17 Meetinghouse Lane is one of the highlights of the Southampton Historical Museum.

AT A GLANCE

Southampton Chamber of Commerce, 76 Main St., 631-283-0402; www.southampton chamber.com or www.hamptonstravel guide.com.

Duck Walk Vineyards, 231 Montauk Hwy., Water Mill, 631-726-7555, www.duckwalk.com.

The wine storage area at Duck Walk Vineyards in Water Mill

Wolffer Estate, 139 Sagg Rd., Sagaponack, 631-537-5106, www.wolffer.com.

Channing Daughters Winery, 1927 Scuttlehole Rd., Bridgehampton, 631-537-7224, www.channingdaughters.com.

Attractions townwide are partially wheelchair accessible and child appropriate.

Did You Know?

Indians called the area Agawam, but English colonists renamed it Southampton — either for the city they originally sailed from or for the earl of Southampton, who sponsored many of the voyages.

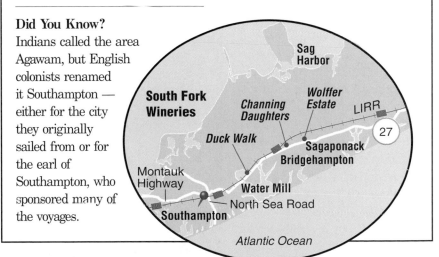

Take the Long Island Expressway to Exit 70 to Route 111 south. Take Route 111 to Sunrise Highway (Route 27). Go east on Sunrise approximately 18 miles to North Sea Road in Tuckahoe. Take North Sea Road south to Main Street, Southampton.

Echoes of Native Culture

BEFORE THE SIMPLE LIFE got so pricey, Southampton was the pristine province of the **Shinnecock Indians** — who lived for aeons off the bountiful earth. Then one June day in 1640 a small band of English settlers sailed in, and the land was soon clogged with trophy houses, the bays with muscle boats. Or so it must have seemed.

The transition actually took a few centuries — during which the Indians shared their farming and fishing expertise (and no doubt more than a few Thanksgiving dinners) with the new neighbors. In return, they saw their boundless territory cut to just over 800 acres by the mid-19th century and, like tribes nationwide, their way of life reduced to film and theme-park fare.

And yet they have endured. "The land is still here, the people are still here and we are still here," says the philosophical narrator of the video "Shared Ground" — one medium used to eloquently tell their saga at the **Shinnecock Nation Cultural Center and Museum**. After a decade of fund-raising, it opened in 2001 on the reservation just east of Southampton College — thanks to a construction grant from the **Mashantucket Pequot Nation**, which runs Connecticut's Foxwoods resort. But the Shinnecocks don't intend to go the casino route, preferring to share their culture via their Labor Day weekend powwow and now their museum.

Newsday Photo / Michael E. Ach

Pride on display at Shinnecock Nation Cultural Center and Museum

246

WHILE YOU'RE THERE

Water Mill Museum, 41 Old Mill Rd., Water Mill, 631-726-4625. **Hours**: 11 a.m. to 5 p.m. Thursday to Monday and 1 to 5 p.m. Sunday late May to late September. **Fee**: $3 adults, $2.50 seniors, children free. This restored working 1644 gristmill features early tools, colonial crafts and hands-on exhibits. Several shows throughout the summer.

Bridge Hampton Historical Society, Montauk Highway and Corwith Avenue, Bridgehampton, 631-537-1088. **Hours**: 11 a.m. to 4 p.m. Tuesday to Saturday June through mid-September, Monday to Friday the rest of the year (closed January and February). **Fee**: $1 adults. The society oversees several early American buildings, including the 1775 Corwith House. Nearby are two barns with turn-of-the-century steam engines and a building with a wheelwright-blacksmith shop and blacksmith demonstrations.

Bridge Gardens Trust, 36 Mitchell Lane, Bridgehampton, 631-537-7440, www.bridgegardens.org. Five acres of plantings from topiary to antique roses. **Hours**: Seasonal.

For a sampling of Southampton restaurants, see Page 295.

The solid front door of the rustic pine log cabin, with its carved deer and elk antler handle, is a preview of the loving handiwork within. Inside, visitors begin "A Walk With the People," from Paleolithic times to the present, via massive murals by Shinnecock artist (and curator) **David Bunn Martine**. Along with his paintings and sculpted totems, exhibits include a replica wigwam, a centuries-old birch bark canoe, mounted animals, implements, jewelry, beadwork, weavings and photos of ancestors and elders.

Some handcrafts are for sale, as are "songs of the struggle" CDs by the reservation's celebrated **Thunder Bird Sisters** (three sisters and a brother, really). The museum may be a work in progress, but long-range plans include an outdoor village, where crafts and rituals will be demonstrated and historical events portrayed. It should be worth the wait. ◆

Shinnecock Nation Cultural Center and Museum, Montauk Highway at West Gate Road, Southampton, 631-287-4923. Call for hours. Fee: $5 adults, $3 seniors and ages 12 and younger.

A Carriage Ride Back To The Slow Lane

A PICTURESQUE village with a thriving university, a revolutionary-era gristmill and a quirky mechanical eagle that flaps its wings on the hour hardly needs anything else to put it on the tourist maps.

Scenic Stony Brook's main attraction, however, is a nine-acre font of Americana that could hold its own on Manhattan's Museum Mile.

The Long Island Museum of American Art, History & Carriages — along with Stony Brook's quaint present-day downtown — owes its existence to the vision and generosity of local tycoon **Ward Melville**, founder of the Thom McAn shoe company. In the 1940s, Melville set out to make Stony Brook a "living Williamsburg," without the gates or admission charge. He rearranged buildings to create a green overlooking the harbor and decreed a genteel architectural style — not quite as widely adopted as he had hoped (the state university, built on land he donated, somehow turned out more New York colossus than Virginia colonial). But his penchant for moving and shaking helped launch the triform museum.

Its art holdings include most existing works of another famous local, **William Sidney Mount**, considered America's first "genre" painter. Mount canvases depicting 19th century country life now share space

Newsday Photo / J. Michael Dombroski

A replica of the old village in the Carriage House section of the museum complex

with varied changing exhibits in the art museum.

The **Bayman's Art Gallery** of the history museum features one of the nation's finest collections of hand-carved wildfowl decoys (as well as occasional carving demonstrations). A simulated former hunting ground near Quogue is evoked via a wall mural, a duck boat — and lots of quacking. Another permanent historical exhibit showcases 15 dollhouse-like miniature rooms, each the size of a computer monitor (no, they're not interactive, but a virtual look wouldn't do justice to the incredible workmanship just in the tiny, stuffed-to-the-rafters antiques shop).

Across the street, a cavernous red barn that looks as if it might be home to a hundred prancing horses holds the renowned collection of vehicles from the carriage era (some hitched to pretty real looking steeds). About 100 carriages, along with such accoutrements as bridles and saddles, are displayed at any time. You'll see dusty buckboards that jounced settlers across the American prairies and gilded coaches that carried European aristocrats; colorfully painted Gypsy wagons and ad-covered commercial carts; one-man carriages and carriages driven by 18th century soccer moms (handling four spirited horses was like managing a station wagon full of kids — but all of them driving, says curator **Merri Ferrell**). There are sleighs and children's carts and fire wagons, including a steam pumper weighing close to five tons — pulled by horses, remember.

You can actually climb aboard a bus that ran between the train station and the old **Stony Brook Hotel**, which once stood on the carriage house site. Occupying the pedestal of honor in the multilevel atrium (how did they get it there?) is a sparkling circa-1880 omnibus last used in 1952 to ferry students around New Hampshire's tony St. Paul's School.

Outside the carriage house are assorted buildings brought from nearby sites: a barn from the Stony Brook farm of Revolutionary War hero **Jedediah Williamson**; a Setauket blacksmith shop whose forge still glows on occasion; the one-room Nassakeag Schoolhouse, where young visitors can squirm through a 19th century lesson during some of the many special museum programs. Near the herb garden is a beaux arts fountain-horse trough that graced New York's Madison Avenue when it was filled with the handsome stretch limos of yesteryear, their grandiose passengers not of a mind to hide behind tinted glass.

If all this makes you yearn to glide along in the slow lane awhile, inexpensive carriage rides are available in the village one day a week spring through fall. ◆

Newsday Photo / J. Michael Dombroski

The children's carriages were pulled by ponies, goats or strong dogs.

AT A GLANCE

The Long Island Museum of American Art, History & Carriages, 1200 Route 25A, Stony Brook, N.Y. 11790, 631-751-0066 or on the Internet at www.longislandmuseum.org. **Hours**: Wednesday to Sunday and most Monday holidays noon to 5 p.m. **Fee**: Adults $4, older than age 60 $3, ages 6 to 17 and college students $2, younger than age 6 free; admission on Monday holidays is $1. Wheelchair accessible. Child appropriate.

Did You Know?
Pint-sized children's carts and carriages were pulled by ponies, goats or sturdy dogs willing to put up with a bunch of children.

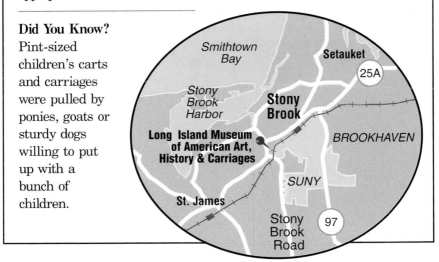

Take the Long Island Expressway to Exit 62 north to Nicolls Road (Route 97). Take Nicolls Road north to Route 25A. Go west on Route 25A about two miles. The museums' parking area is on the right.

◆

Grinding for Their Daily Bread

THE VILLAGE GRISTMILL was more than a dusty flour factory. It also was an early American social club where townsfolk could escape their own daily grind.

Farmers transacted business, traded news and gossiped there while waiting for their grain to be processed. Children were endlessly entertained by the roaring machinery and splashing waterwheel. One 19th century Stony Brook miller notched up the festive atmosphere by playing Gypsy violin music for the regular crowd.

Stony Brook's first gristmill, built in 1699, was washed away by a flood. But its 1751 replacement, now listed on the **National Register of Historic Places**, has been busy since colonial times, when British troops regularly raided it for provisions during the Revolutionary War. When it was retired in the 1940s, it was shipping freshly ground wheat germ daily to health conscious customers in 42 states.

Today's visitors get a demonstration of the ingeniously simple milling process from top to bottom (though it doesn't strictly progress from the fourth floor attic down to the ground). Guides are a costumed "miller" and sometimes his young apprentice, known as a dusty (you quickly see why). Briefly, grain is air-cleaned, ground between two 2,000-pound millstones into flour or meal (called grist), then sifted. In the backyard, you can watch the rocky stream that gave the village its name tumble over the waterwheel — ever turning, turning, turning to provide the power.

Newsday Photo / Bill Davis

The Stony Brook Grist Mill, built in 1751 to replace a 1699 mill destroyed by a flood, is on the National Register of Historic Places.

The **Stony Brook Grist Mill** twice did double duty: in the 1840s as a sawmill and in the 1880s as a winery that produced a precocious little chardonnay (as far as 21st century researchers can tell). Grapes still grow wild nearby, seeded from the vineyard that once graced an island in the mill pond.

The gristmill's country store doesn't sell wine, or products made on-site. But you can buy stone-ground wheat flour, cornmeal and pancake mix as well as jams, syrups, cookbooks, antique kitchen utensils and corncob dolls. And, of course, cracked corn to feed the ducks in

WHILE YOU'RE THERE

Stony Brook Village walking tours. **Hours**: Every Wednesday in July and August. Meet at 2:50 p.m. at the post office, where the mechanical eagle over the door flaps its wings on the

Newsday Photo / J. Michael Dombroski

hour from 8 a.m. to 8 p.m. daily (it's synchronized to the clock on the cupola, so adjust your watch). **Fee**: $2 adults, $1 under age 12.

The Three Village Garden Club Exchange, in a reconstructed building that combined the former firehouse and the "Old Stone Jug," offers two floors of antiques and collectibles (signs warn "if you break it, you buy it," but even a bull would rarely be out more than $25 in this modest china shop, where more fragile treasures are in cases). **Hours**: Noon to 5 p.m. Monday to Sunday; 631-751-0560.

Harbor-front **Hercules Pavilion** houses a Hercules figurehead and anchor from the Ohio, the first ship launched from the Brooklyn Navy Yard (1820), plus a whaleboat thought to be the only remaining artifact from the Charles Hall 1872 Antarctic expedition.

While you're checking out the several dozen specialty shops of **Stony Brook Village Center,** notice the little gabled, brown-shingled church — **All Souls Episcopal** — on the hill next to Market Square. It was built in 1889 from a design by the noted architect Stanford White, with its roof lines suggesting praying hands.

Naturalist-guided cruises on the 35-passenger **Discovery,** 631-751-2244. **Hours**: May through October. Call for schedule. **Fee**: $15 adults, $9 under age 12.

For a sampling of Stony Brook area restaurants, see Page 296.

T. Bayles Minuse Mill Pond Park. There are frequent special events, too. ◆

Stony Brook Grist Mill, Harbor Road (off Main Street), 631-751-2244. Hours: Noon to 4:30 pm Wednesday to Sunday June through August, noon to 4:30 p.m. Saturday and Sunday in April, May and September through the beginning of December. Fee: $2 adults, $1 under age 12.

SUFFOLK COUNTY VANDERBILT MUSEUM

A Mansion Untarnished By Time

I T'S SURELY the only house on Long Island that's been able to boast a 3,000-year-old Egyptian mummy reposing in the guest wing and a 238-seat planetarium in the backyard.

But rare relics from ancient cultures and indoor sky shows complete with thundershowers are just two highlights of **William K. Vanderbilt II**'s Centerport estate — which began as a seven-room bachelor retreat in 1910 and evolved into the distinguished **Suffolk County Vanderbilt Museum**.

Maybe you can't actually live the Vanderbilt life there, but you can indulge your inner rich kid by touring the exquisite (yet invitingly homey) mansion and enjoying seasonal entertainment on the 40-acre grounds overlooking Northport Harbor. Summer visitors are made to feel all the more like illustrious guests of the 1930s by costumed actors who play Vanderbilt's family members, service staff and socialite friends.

This Vanderbilt was a great-grandson of shipping and railroad titan **Cornelius Vanderbilt**, who is credited with creating the family fortune (he made it, Willie K. spent it, visitors are told). Willie eventually rose to vice president of flagship New York Central Railroad, but perhaps saw more of the two giant iron eagles that once perched atop its Grand Central Terminal headquarters after they came to rest on the gateposts of his hilltop

254

An arcade and garden at the Spanish Revival-style Vanderbilt Museum

estate, called **Eagle's Nest**. Eclectic interests from art and anthropology to marine science and automotive technology occupied most of his time, but also provided an outlet for philanthropy. When Long Island roads got too crowded to safely handle his passion for grand-prix racing, he built a private parkway from Queens to Suffolk for his international **Vanderbilt Cup** competitions, later opening it to all cars — albeit with a $1 toll (you can drive today on a short stretch, now called Vanderbilt Motor Parkway, for free).

During countless scientific expeditions around the world (his 265-foot yacht carried a seaplane and a crew of 50, including a taxidermist), he discovered many new marine species — which he was thus entitled to name (his beloved second wife, **Rosamond**, likely realized it was an honor to have a crab named after her). Once a week, he invited the public onto his estate to admire the brightly painted specimens he displayed in what is now called the **Hall of Fishes**.

From the get-go, Vanderbilt declared his intent to turn his home into a museum, which allowed him to import antiquities that would be off-limits to a casual collector. He scanned the globe with a connoisseur's eye and a gentleman's sensibilities — more than once swapping some of the spiffy American power shirts in his trunk for indigenous authority symbols from Polynesia.

The Egyptian mummy, which arrived in the 1930s (and is now complete with CAT scans), was due to move from a guest room to the laboratory in the newly refurbished curator's cottage, where other artifacts are displayed.

Thousands of carefully chosen acquisitions artfully fill the house, which by 1936 had grown into the present 24-room balconied Spanish Revival villa. The final addition was a **Memorial Wing** devoted to safari trophies and other mementos of his son, **William K. Vanderbilt III**, who died in a 1933 car accident at age 26. (The planetarium was added 27 years after Vanderbilt's own death in 1944.) You'll see furnishings and objets d'art from every continent, as well as hand-carved woodwork (often adorned with scallop shells, a sailors' good luck charm). Don't let the water view from the window-walled master suite keep you from noticing a couple of ingenious curiosities: the toe-tester spigot in the shower stall and the Georgian chair with a seat-back shelf (whose purpose you'd never guess if a docent didn't tell you). **Vanderbilt's 1928 Lincoln** touring car is on display on a basement turntable that was used by trucks making deliveries to the mansion so they didn't have to back down the narrow, winding driveway.

Small touches throughout the living quarters (a bottle of Champagne chilling by the bathtub, "dinner" on the dining room sideboard ready to be served) make the house look as if its residents are apt to walk in at any moment. It's hard to know if this was done solely to enhance the scrupulous historical authenticity — or also to deter visitors who might otherwise want to move right in. ◆

Newsday Photo / Bill Davis

The table set, visitors might wonder if a Vanderbilt will sit down to dinner.

AT A GLANCE

Suffolk County Vanderbilt Museum: 180 Little Neck Rd., Centerport, 631-854-5555 or on the Internet at www.vanderbilt museum.org. **Hours**: Noon to 5 p.m. Tuesday to Sunday. **Fee**: Planetarium show or mansion tour / grounds pass $8 adults, $6 over 60 and students ages 12 through college, $4 under age 12, free under age 2; combination tickets $11, $9 and $7. **Theater**: Call for programs and prices. **Museum**: Partially wheelchair accessible. Partially child appropriate. **Planetarium**: Wheelchair accessible. Child appropriate.

Did You Know? Vanderbilt's German-built yacht, Alva, became the Plymouth when he gave it to the U.S. Navy in World War II. It was sunk by a Nazi U-boat in 1943.

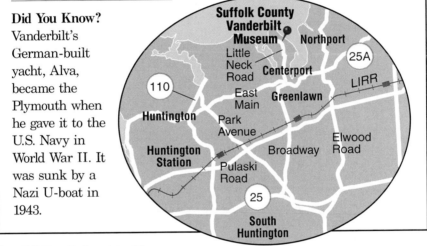

From Exit 49 on the Long Island Expressway, go north on Route 110 to Route 25A in Huntington. Take 25A east about three miles to Centerport. Take Little Neck Road north about 1½ to 2 miles. Vanderbilt Museum and Planetarium is on right.

Ceiling of the Stars

A S YOU SETTLE into your seat, twilight slowly envelops a virtual neighborhood that looks a lot like yours. But here, without the alpenglow of streetlights, the night sky deepens to an ebon velvet that suddenly sparkles with a zillion stars.

Who knew all this is always overhead?

When the lights dim in the **Vanderbilt Museum's planetarium**, you're in for multiple surprises. The universe projected on its dome is always wondrous, yet never scary (though its thunderstorms will spatter a few drops of rain on your cheeks and its lightning flashes may prompt the youngest family members to momentarily reach for a reassuring hand).

Asteroids and constellations whirl before your eyes, until it almost seems you're rocketing toward them. Just glance down if you're nervous during blast-

Newsday Photo / Bill Davis

Visitors to the Vanderbilt Planetarium sit back and travel the universe.

off. But don't worry; this space odyssey is perfectly safe (at worst, you might lose a shoe in the dark). All that's missing from your brief journey through the solar system is that elusive pot of gold at the end of the rainbow. Naturally, you can wish on a star.

Three public sky shows alternate at any given time, and all are tweaked periodically (school groups see others, tailored to age). The newest public show is "Mars Adventure."

Want to delve deeper into the mysteries of the galaxies? There are numerous lectures, courses and other programs. You can also call staffers with questions relating to astronomy (they began re-evaluating Pluto as soon as New York City's **Hayden Planetarium** demoted it from a planet to a comet). The planetarium was added to the estate in 1971, when the United States was obsessed with the space race, and is now considered among the nation's dozen largest and best equipped. The sliding-roof observatory, open on clear evenings when there are sky shows, includes the largest public-use telescope in the New York metro area. Oh, and don't overlook the gift shop, which stocks everything from geodes to Astronaut Ice Cream (a surprisingly tasty freeze-dried Neapolitan chunk). ◆

Suffolk County Vanderbilt Museum Planetarium's public sky shows run on weekends, vacation weeks and some Monday holidays during the school year. They're also offered weekdays (except Monday) during the summer. Weekday and Saturday shows normally are 11 a.m., 1 and 3 p.m.; Sunday's are noon, 1:30 and 3 p.m.

WHILE YOU'RE THERE

Walt Whitman, considered one of America's greatest poets, may have left his Huntington birthplace when he was 4, but he returned to Long Island for inspiration throughout his life. Though engulfed by commercial spillover from the Route 110 corridor, the **Walt Whitman Birthplace State Historic Site** at 246 Old Walt Whitman Rd., Huntington Station, offers insights into the author. Tours are offered of the restored circa-1819 farmhouse and original manuscripts and memorabilia are featured at an interpretive center-bookstore. **Hours:** Open daily except holidays (also closed Tuesdays in summer, Mondays and Tuesdays in winter). **Fee:** Adults $3, seniors and students $2, ages 7 to 12 $1. For other information, call 631-427-5240.

In nearby 854-acre **West Hills County Park**, off Sweet Hollow Road, visitors can follow wooded trails Whitman may have trod to the top of Jaynes Hill, Long Island's highest point (parking fee summer weekends and holidays; 631-854-4423).

Suydam Homestead Barn Museum, 1 Fort Salonga Rd. (Rte. 25A), Centerport, 631-754-1180 or http://gcha.suffolk.lib.ny.us. **Hours:** 1 to 4 p.m. Sundays June through October. **Fee:** Donations. The circa-1730 homestead has a display of archeological finds, local china and tools.

For a sampling of nearby restaurants, see Huntington on Page 287 and Northport-Greenlawn on Page 289.

Newsday Photo / Bill Davis

Poet Walt Whitman's birthplace in Huntington

SUNKEN MEADOW STATE PARK

Pastoral Views, And Plenty To Do

THE LANGUOROUS entrance parkway rises so imperceptibly that first-time visitors generally blink in amazement when it suddenly unfurls a panorama of French-blue water lapping the hazy distant horizon.

This knockout vista across Long Island Sound and Smithtown Bay is just one of **Sunken Meadow State Park's** many surprises — not the least of which is its dramatically diverse topography.

Water views are never far from sight anywhere within the 1,266 acres of sandy bluffs, glacial cliffs, dense woods, salt marshes and brackish creeks (the park's eastern border is the delta of the **Nissequogue River**). There's hiking, biking and horseback riding (but no rental horses are available — it's strictly BYOH), windsurfing, beach and creek fishing, picnicking and golf, plus ball fields and basketball courts on the low "sunken" meadowlands. When there's snow, you'll always see some cross-country skiers.

An arcaded art deco bathhouse, now office space, is still the imposing entrance to a beach built when bathers had to shed their soggy wool swimsuits before climbing into their Model Ts' upholstered rumble seats. While Sunken Meadow's soft sand is never as mobbed as the South Shore's ocean strands, it attracts more than 35,000 visitors each day some weekends during swimming season (Memorial Day to a week after Labor Day, when life-

Taking a stroll on the boardwalk at Sunken Meadow State Park

guards are on duty). You'll find everyone from the sand castle-building toddler brigade (there's some children's entertainment) to daydreamers who in their minds board every passing sailboat and wisp of cloud.

The No. 1 attraction of this year-round park, however, is the .75-mile boardwalk — often more crowded on an unexpectedly mild February Sunday than on the hottest day in July. Winter visitors, it seems, savor the chance to shake off cabin fever there, whereas beachgoers just scoot across it to the water.

Strollers (both the hand-holding and baby-conveying variety) as well as stopwatch-checking exercise buffs (from headset-wearing joggers to waddling power walkers) amicably share the boards. Mid-September to mid-May, bicyclists are allowed, too (look for one perched on a shoulder-high unicycle). Roller-skating and skateboarding are banned only because they damage the planks — which are smooth except for a scattering of clam and mussel shells smashed by hungry gulls preparing hors d'oeuvres. There are also two seasonal refreshment stands and a pavilion for catered events.

Hikers, runners, bird-watchers and lovers seeking a few moments of seclusion share the park's six miles of trails. **The Long Island Greenbelt** wends its way across a footbridge at the southwest corner of Parking Field 3, follows the Nissequogue and terminates in Heckscher State Park 34 miles away on Great South Bay. (The Greenbelt organization also has outlined a four-mile loop from Sunken Meadow that it calls the "low road at low tide route.") Park trails include overlapping 5K (3.1-mile) and five-mile courses used for cross-country races and each June for the state parks **Summer Run Series** 10K (6.2-mile) race.

A mile down a pastoral two-lane road past Parking Field 2 (reserved for off-season model airplane flying and spring and fall air shows) is **Sunken Meadow Golf Course**. It's actually three nine-hole links (the Blue, Green and Red) set up as one nine-hole and one 18-hole course in rotation every three days. There's also a 500-yard driving range, a putting green, a clubhouse — and a public tournament each spring.

When the park opened in 1928 (a year before Jones Beach), it was among the first retreats for city folks created by the visionary **Robert Moses** as head of the Long Island State Parks Commission. Half a century later, the state parks department decided to rename it in honor of former **Gov. Al Smith**, who had supported Moses' myriad projects. But while it's officially Gov. Alfred E. Smith-Sunken Meadow State Park, it remains simply Sunken Meadow to its countless fans. ◆

Newsday Photo / David L. Pokress

Sunken Meadow Golf Course features three nine-hole links.

AT A GLANCE

Sunken Meadow State Park, northern terminus of Sunken Meadow State Parkway, Kings Park, 631-269-4333 for general information and permits, 631-269-5351 for golf. **Hours**: Open daily sunrise to sunset. **Fee**: $7 per vehicle during swimming season, $5 weekends and holidays in spring and fall. No pets. Golf fees vary. Partially wheelchair accessible. Child appropriate.

Did You Know?
During a 12-month span, more than 150 species of birds have been spotted in the park (which can provide a free bird list on request).

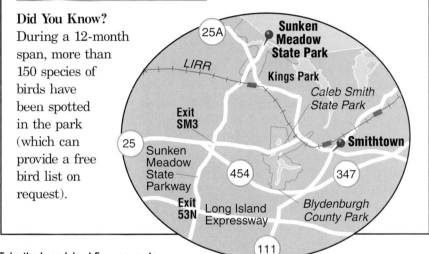

Take the Long Island Expressway to Exit 53 north to Sunken Meadow Parkway. Take the parkway north about six miles to the park. *For Smithtown,* return south on Sunken Meadow Parkway to Exit SM3. Go east on Route 25 about two to three miles to Smithtown.

Double the Parks, Double the Fun

T HE SMITH FAMILY understandably pops up everywhere in Smithtown — including in two adjacent two-part parks.

Caleb Smith State Park Preserve's 543 acres date to the original tract acquired in 1663 by town founder **Richard "Bull" Smith**, who legend says made a pact with the Indians that he could keep all the land he circled in a day riding on his bull (history favors a more traditional transaction). Two hundred years later, his descendant, Caleb Smith, sold parts of the property to a sportsmen's club; a century after that, it became a state park, bisected by Jericho Turnpike (Route 25).

The northern portion has nature trails (including a half-mile, all-access loop and a section of the Long Island Greenbelt) plus a small nature museum in the expanded original **1751 Caleb Smith farmhouse**. Besides bird walks and nature programs, a junior anglers' area is open from April through October. The wilder section south of the highway offers fly-fishing along the Nissequogue River April through mid-October (by permit and reservation).

Newsday Photos / Tony Jerome

A busy dock at Blydenburgh County Park

Adjoining 588-acre **Blydenburgh County Park** isn't bisected by a highway but by 120-acre Stump Pond (also called Mill or New Mill Pond), created by damming the Nissequogue River. The park's southern part has picnic and camping areas, rowboat rentals, hiking and freshwater fishing (no swimming); the northern segment embraces a 10-acre National Register Historic District. You can walk between the two, but it's about four miles round-trip, so most visitors drive (which requires leaving and re-entering the park). The historic district includes two sites under restoration: a circa-1800 commercial hub established by the Smith and Blydenburgh families and a farm complex from the later Victorian period; the **Blydenburgh-Weld House** is headquarters for the Long Island Greenbelt Trail Conference, which offers tours Saturdays at 1 p.m. ◆

Caleb Smith State Park Preserve, Jericho Turnpike, Smithtown, 631-265-1054. Hours: 8 a.m. to sunset Tuesday to Sunday April to September, Wednesday to Sunday October to March. Fee: Parking $5 daily April to Labor Day, then weekends and holidays. Blydenburgh County Park, 631-854-3713 (631-360-0753 for historic district tours). Hours: Year-round dawn to dusk Memorial Day through Labor Day. Fee: Parking $2 Suffolk residents, $4 others. Historic district entrance at end of New Mill Road (off Jericho Turnpike or Brookside Drive); recreation entrance is off Veterans Highway (Route 454).

Autumnal colors paint the landscape at Caleb Smith State Park Preserve.

Newsday Photo / Bill Davis

Part of the Long Island Greenbelt Trail at Nissequogue River State Park

WHILE YOU'RE THERE

Caleb Smith House, 5 North Country Rd. (Route 25A), Smithtown, 631-265-6768. **Hours:** 9 a.m. to 4 p.m. Monday to Friday, noon to 4 p.m., Saturday. **Fee:** Donation. It was actually Caleb Smith II who lived in this house that was built in the 18th century and expanded in 1819, now home to the Smithtown Historical Society.

Nissequogue River State Park, St. Johnland Road, Kings Park, 631-269-4927. **Hours:** Open year-round sunrise to sunset. **Fee:** $5 per car daily Memorial Day through Labor Day, weekends and holidays late spring and early fall. Fishing, hiking and birding are offered on this 153-acre park situated on the Nissequogue River.

Arthur Kuntz County Park, off Landing Avenue, Smithtown, 631-854-4949. **Hours:** Year-round. **Fee:** Free. This 97-acre Suffolk County park, pierced by the Greenbelt Trail, is popular with hikers and bird-watchers. It offers exciting views of Nissequogue River Valley.

David Weld Sanctuary, Boney Lane, Nissequogue, 631-367-3225. The 125-acre horseshoe-shaped Nature Conservancy property includes diverse habitats and 1,800 feet of beach frontage on Long Island Sound. Marked trails open to walkers dawn to dusk.

For a sampling of Smithtown-Kings Park restaurants, see Page 295.

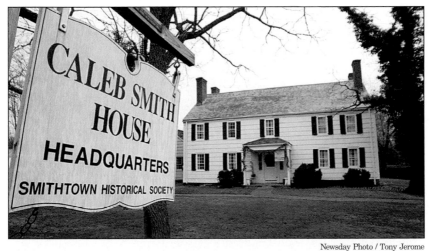

Newsday Photo / Tony Jerome

Caleb Smith House in Smithtown, former home of Caleb Smith II

DISCOVER LONG ISLAND

Keeping Things Shipshape At Kings Point

T FIRST GLANCE, it could be any college campus: stately buildings, tree-shaded walks, go-team graffiti (just a tad). Then a drumroll sounds, and the main square is suddenly filled with knife-crease khakis and jaunty garrison caps. The uniformed students fall into formations and march smartly off — in their daydreams to sail the seven seas, but for now only to lunch.

"Noon Muster" (actually held at 11:30 a.m. — don't ask) is part of the daily pomp and pageantry at the **U.S. Merchant Marine Academy at Kings Point**, probably the least heralded of the nation's five federal service academies. That's because Kings Point trains officers for civilian careers on lackluster commercial ships while the others turn out military leaders bound for glory (West Point for the Army; Annapolis, Md., the Navy; Colorado Springs, the Air Force; and New London, Conn., the Coast Guard).

But there are many similarities. Like the others, Kings Point is open free to the public most days (your taxes pay for it, after all). And it adheres to the same strict spit-and-polish standards. Stroll the 82-acre campus on **Long Island Sound** and you'll often be greeted with a good morning, ma'am or sir (yes, even budding freighter captains get etiquette lessons).

Indeed, each of Kings Point's 900 midshipmen (the term also is applied

268

Newsday Photo / Ken Spencer

Midshipmen take lunch in the Merchant Marine Academy's dining hall.

to the 10 percent who are women) graduates with a commission in the **Armed Forces Reserves** as well as a bachelor of science degree and a merchant marine license (to repay the government for their education, grads must do stints of both maritime and reserve service, and many also serve during times of war). But what benefits. By the end of their junior year, cadets have visited 18 or 19 countries on academy training vessels. And you thought all young adventurers joined the *Navy* to see the world.

Brochures available at the main gate outline a short self-guided walking tour of the grounds, which are dotted with memorials and nautical bric-a-brac such as giant anchors. The campus also offers the fine **American Merchant Marine Museum,** which will fill you to the gunwales with ships, ships and more ships.

Like many Long Island institutions, this one grew out of a wealthy man's estate — in this case that of auto magnate **Walter P. Chrysler.** The white stucco, chateau-style mansion, with its rich wood paneling and hand-painted ceilings, provided class and dormitory space during the early months of the academy — which also were the early months of World War II (most of today's campus was built then, all within 18 months). Amid historical nautica in the lobby of the house, now called **Wiley Hall,** don't miss the framed

269

copy of "The First Lady March," written by the academy music director, Capt. Kenneth Force, for Hillary Rodham Clinton and performed when she spoke at graduation in 2000.

In front of Wiley Hall, near a replica of the **battleship Missouri** plaque commemorating the end of World War II, is the tallest flagpole east of the Mississippi (whose dizzying 174-foot height daunts mischief by even the most athletic plebes). On the water side of the mansion, visit the interfaith chapel to admire its three-sided altar, which rotates on a turntable.

The maritime museum — easy to spot by the 19-ton troopship propeller out front — is in another former private home. This one belonged to **William S. Barstow**, an inventor colleague of Thomas Edison. The first floor is filled with ship models (some in bottles, others as big as a canoe), ship posters and paintings, memorials and memorabilia. A prized possession is a double steering wheel from the **Constitution** — lost during the War of 1812 battle that earned the ship its nickname, "Old Ironsides." How could it lose its wheel and still win the skirmish? You'll find out.

If all this sea duty works up an appetite, you can stop for a snack at the **Seafarer** cafe on the ground floor of Delano Hall, then browse the "ship's store" for logo souvenirs. A pair of boat shoes, perhaps? ◆

A WWII troopship propeller at the American Merchant Marine Museum

AT A GLANCE

U.S. **Merchant Marine Academy** and **American Merchant Marine Museum**, Kings Point, 516-773-5000 (museum 516-773-5515) or at www.usmma.edu. **Hours:** Academy 9 a.m. to 5 p.m. daily, museum Tuesday to Friday 10 a.m. to 3 p.m., Saturday and Sunday 1 to 4:30 p.m.; campus and museum closed in July during basic training, and on federal holidays. **Fee:** Academy free; $1 donation requested at museum. Partially wheelchair accessible. Child appropriate.

Did You Know? The U.S. Merchant Marine Academy began admitting women in 1974, two years before the other federal service academies.

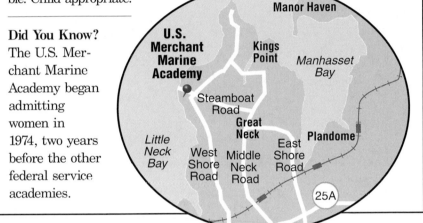

Take the Long Island Expressway to Exit 33 north to Community Drive. After crossing Northern Boulevard, Community Drive becomes East Shore Road and then becomes Hicks Lane. Turn right off Hicks Lane at Middle Neck Road. Go one light to Steamboat Road. Make a left onto Steamboat Road. Continue one mile to the academy.

◆

Learning Through Doing

WHAT DO CRABS, butterflies and bubbles have in common? They're all things parents and kids can learn about together in Saturday family programs at the **Science Museum of Long Island** in Plandome.

This isn't a traditional museum filled with displays you merely look at; it's a hands-on science activity center in a 1911 mansion now filled with classrooms at the **Leeds Pond Preserve.** Seven days a week there's some-

Newsday Photo / Karen Wiles Stabile

Educators at the Science Museum of Long Island hold a chicken and a duckling hatched there two weeks earlier.

thing going on — experiments, demonstrations, field trips; about 130 annual programs in all regularly attract kids, adults and families from well beyond the Island. The museum also sponsors research and informational expeditions to such exotic destinations as Cuba and Easter Island for adults — who sometimes have an opportunity to assist world-famous scientists, many of whom also have lectured at the museum.

The 36-acre preserve overlooking **Manhasset Bay** provides lots of outdoor learning opportunities, too. The property includes a forest, salt marsh, brackish pond, freshwater stream and estuarine beachfront. In conjunction with Adelphi University, there's an ongoing archeological dig of a major American Indian dwelling site near the water's edge. ◆

Science Museum of Long Island, 1526 N. Plandome Rd., Plandome, 516-627-9400 or on the Internet at http://ourworld.compuserve.com/home pages/smli. Preregistration required for programs.

WHILE YOU'RE THERE

The restored early 18th century **Saddle Rock Grist Mill** on Grist Mill Lane (off West Shore Drive) in Great Neck, 516-571-7900. **Hours**: Open for tours May to October, Sundays 1 to 5 p.m. **Fee**: Free.

Temple Beth Shalom Judaica Museum, 401 Roslyn Rd., Roslyn Heights, 516-621-2288. **Hours**: 9 a.m. to 5 p.m. Monday to Thursday, 9 a.m. to 1 p.m. Friday and Sunday; call for Sabbath hours. **Fee**: Free. The museum provides a picture of the heritage of Jewish people around the world through permanent and changing exhibits of ritual objects, photographs and paintings.

Cedarmere, Bryant Avenue, Roslyn Harbor, 516-571-8130. **Hours**: 1 to 5 p.m. Saturday and Sunday, April through early November. The Long Island home of William Cullen Bryant, 19th century poet and newspaper editor, features exhibits, furnished rooms, restored gardens, pond and mill.

For a list of nearby restaurants, see Great Neck on Page 285.

Cedarmere, home of poet William Cullen Bryant, a leading poet and publisher who settled in Roslyn in 1843

Newsday Photo / Dick Kraus

WILLIAM FLOYD ESTATE
AND THE MANOR OF ST. GEORGE

Proud Homes
Swept Up
In a Revolution

HE SIGNED the document that launched our nation. But how many beachgoers who ply the parkway named for him know that was William Floyd's claim to fame — or that the colonial estate where he was born is an intriguing mainland part of **Fire Island National Seashore**?

On the eve of the American Revolution, Floyd was more a model conservative landowner than a rebellious patriot. But as a Suffolk County representative to the Continental Congress in Philadelphia, he felt compelled to add his name to the Declaration of Independence. This didn't sit well with local British loyalists, who occupied his farm in what is now **Mastic Beach** and forced his family into a seven-year exile in Connecticut (during which his wife died).

When he finally returned to Suffolk with his children and found his estate largely trashed, he embarked on a remodeling project to make the house more suitable for entertaining prominent new best friends such as founding fathers Thomas Jefferson and James Madison (soon there also were a new wife and more offspring).

From 1718 to 1976 eight generations of Floyds managed the property, adapt-

Newsday Photo / Dick Kraus

The family of revolutionary patriot William Floyd occupied this house for more than 250 years before donating it to the National Park Service.

ing it to their changing needs, so today's visitors see a continuum of features and furnishings. Artifacts traceable to **William Floyd**, who was born there in 1734, include paintings, a writing desk and a traveling liquor chest.

In the custom of those unenlightened times, Floyd operated his farm plantation-style, and white wooden crosses outside its cemetery fence still mark the graves of slave laborers. (Floyd is buried upstate, where he moved his family after the turn of the 19th century to start a new life on the "frontier.")

Twentieth-century Floyds started scaling back their involvement in the estate, eventually using it mainly for summer recreation and winter hunting. In 1975, the last family member donated it to the **National Park Service**, which tries to preserve historical accuracy.

Except for the entrance drive, roads on the property are dirt and cars are parked out of view of the house. There are a few picnic tables, but don't even think of spreading a blanket on the front lawn. (Can you imagine history buffs wanting spiky pink hair and cell phones in their snapshots of a prized colonial landmark?)

While the Floyd house grew in size over the years, the grounds shrank from the original 4,400 acres that extended north to the area now occupied by **Brookhaven National Laboratory**. But the 613 remaining acres offer 8.5 miles of trails through salt marsh, fields and second-growth forest bursting with birds and deer.

The Floyd estate had been part of a tract of tens of thousands of acres England's King William and Queen Mary granted in 1693 to a favorite court attache known as Tangier Smith, who was mayor of Morocco's capital when it was an English "royal city." On Long Island, Smith became the first lord of the **Manor of St. George**, now a park and museum on 127 acres overlooking Bellport Bay, a few miles west of Floyd's home.

In 1776, the British occupied the Manor of St. George and turned it into Fort St. George. But American forces recaptured it in 1780 via a surprise 4 a.m. attack that lasted just 10 minutes, and **Tangier Smith**'s descendants soon returned. The dual history is still acknowledged, however, by the flag that flies outside: a precursor to the Stars and Stripes with a mini-Union Jack in place of the field of stars.

Rooms shown on tours of the two-story house include the original kitchen with its open-hearth fireplace and regal-width wallboards (the mark of a royally sanctioned project, built from the biggest trees).

Most manor residents over almost three centuries are buried in the cemetery on the grounds. But Tangier, who died there in 1705, lies in a Smith family plot in Setauket — like Floyd, ever the sojourner. ◆

Newsday Photo / Dick Kraus

The Manor of St. George sits on land established by a royal patent in 1693.

AT A GLANCE

William Floyd Estate, 245 Park Dr., Mastic Beach, 631-399-2030. **Hours**: House tours 11 a.m. to 4:30 p.m. Friday to Sunday and major holidays from late May through October; grounds open January through October Saturday and Sunday 9 a.m. to dusk. **Fee**: Free. Child appropriate. **The Manor of St. George**, Neighborhood Road and William Floyd Parkway, Shirley, 631-281-5034. **Hours**: Wednesday to Sunday, May through late October, house tours 10 a.m. to 4 p.m., grounds open 9 a.m. to 5 p.m. (no dogs or picnicking). **Fee**: Free. Partially child appropriate. Both sites are partially wheelchair accessible.

Newsday Photo / Dick Kraus

A portrait of William Floyd on the wall of the main hall at the William Floyd Estate.

Did You Know? Ceiling height in colonial homes had more to do with a family's status than stature — invariably lower in servants' quarters.

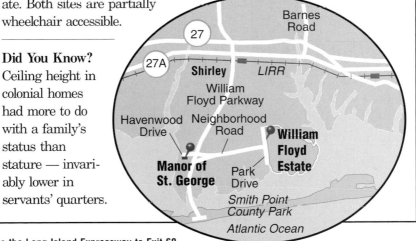

Take the Long Island Expressway to Exit 68.
Go south on the William Floyd Parkway. After passing Sunrise Highway (Route 27), continue for three miles. Make a left onto Havenwood Drive. This road will quickly become Neighborhood Road. Take Neighborhood Road for two miles and make a left onto Park Drive. William Floyd Estate is four blocks on the right. Get directions there to the Manor of St. George.

Where Energy Matters

FOR MORE than half a century, scientists from around the world have pursued cutting-edge research at **Brookhaven National Laboratory**. But to its suburban neighbors, the mini-city deep within 5,300 acres of the Upton woods essentially remains a benign mystery (though some no doubt have wondered if ongoing experiments to recreate the Big Bang could possibly beam Long Island back to the beginning of time).

To dispel any such notions — and boast about discoveries ranging from more energy efficient buildings to improved medical diagnostics — the lab annually invites the public in for eight free, family-friendly Summer Sunday Tours. Each daylong open house focuses on a different field of science.

Brookhaven National Laboratory

An exhibit at Brookhaven National Lab explores the physics of gravity.

The nonprofit lab (now funded by the U.S. Energy Department) was founded in 1947 by a handful of eastern Ivy League universities to study peaceful uses of the atom. It was carved out of the Army's old Camp Upton — where a World War I sergeant named Irving Berlin wrote the musical "Yip, Yip Yaphank," featuring the hit "Oh, How I Hate to Get Up in the Morning" (he returned to write "This Is the Army" when the post was reactivated during World War II). A small museum recalling military days is open during tours.

Since then, Brookhaven's experiments have yielded four Nobel Prizes in physics as well as breakthroughs in research on Lyme disease, salt-hypertension, diabetes and brain-centered problems such as drug addiction, schizophrenia, Alzheimer's and Parkinson's. An ongoing NASA project is studying likely effects of travel to Mars. The lab's most used facility is the **National Synchrotron Light Source**, whose complex functions are often the focus of a summer tour day.

But Brookhaven's current pride and joy is a Relativistic Heavy Ion Collider, billed as the world's newest and biggest particle accelerator (the top of

WHILE YOU'RE THERE

Smith Point County Park, southern end of William Floyd Parkway, 631-852-1316. **Fee**: Parking $10 late May to early September ($5 for Suffolk Green Key holders); free admission for seniors and disabled persons on weekdays, except holidays. The 2,290-acre park offers a five-mile protected beach (late May to early September), saltwater fishing, hunting (license required), picnicking, camping.

Otis Pike Fire Island High Dune Wilderness Area, southern end of William Floyd Parkway, 631-281-3010. **Hours**: Visitor center open 8 a.m. to 4:30 p.m. Memorial Day through Labor Day (varying hours rest of the year); disabled parking only (others must park at adjacent Smith Point County Park); exhibits, beach camping and nature trails.

Wertheim National Wildlife Refuge in Shirley (take Smith Road south off Montauk Highway for a quarter mile to entrance road on the right), 631-286-0485. **Hours**: 8 a.m. to 4:30 p.m. year-round. **Fee**: Free. The 2,550-acre refuge, bisected by Carmans River, has two hiking trails accessible from the parking lot, one only by boat (kayak, canoe rentals nearby), wildlife viewing, fishing (from boats only), environmental education, cross-country skiing.

Suffolk County Farm and Education Center, 129 Yaphank Ave., Yaphank, 631-852-4600. **Hours**: 9 a.m. to 3 p.m. daily. **Fee**: Free for self-guided tours; fee for special programs (which require registration). This century-old working farm operated by the Cornell Cooperative Extension features farm animals, hayrides, horse shows, picnic areas, family and senior programs.

For a sampling of nearby restaurants, see Bayport-Patchogue area on Page 280 and Hampton Bays-Yaphank area on Page 286.

its 2.4-mile-long, doughnut-shaped tunnel shows up on satellite photos like a giant backyard mole run). About 1,000 international scientists have collaborated on the design and operation of the collider, which they hope will replicate conditions thought to have existed at the beginning of the universe. It may be on the Summer Sunday Tour schedule, but visitors naturally won't be allowed to witness the creation. ◆

Brookhaven National Laboratory, William Floyd Parkway, Upton, 631-344-2651, www.bnl.gov.

Restaurants

WHILE YOU'RE OUT discovering Long Island, you might get hungry. Here are some restaurants in various price ranges recommended by Newsday dining critics Peter M. Gianotti and Joan Reminick. The eateries are organized by location.

AQUEBOGUE, SOUTHOLD

Jamesport Country Kitchen, 1601 Main Rd., Southold, 631-722-3537. Moderate. Jamesport Country Kitchen exemplifies what's best about the North Fork. In a setting defined by flowered wallpaper and lace curtains, chef-owner Matthew Kar draws on the bounty of local fishermen and farmers. His preparations, such as his signature salmon cakes, are simple and skillful, allowing the ingredients to speak for themselves.

Modern Snack Bar, Main Road (Route 25), Aquebogue, 631-722-3655. Inexpensive. A local classic that has been serving homey food since Truman was president. Roast turkey, meat loaf, lobster, bay scallops, roast duckling, sauerbraten, and a quart of the house's mashed turnips to go. Tasty fruit pies, too.

BAYPORT, PATCHOGUE AREA

Louis XVI, 600 S. Ocean Ave., Patchogue, 631-654-8970. Very expensive. Excellent French restaurant with water view. Elaborate decor, first-rate service. In season, there's an abundance of truffles. Potato-crusted snapper and rack of lamb are superior. The desserts are showstoppers year-round, particularly the apropos Marie Antoinette Doll with its spun-sugar headpiece.

Thai Lemongrass Grill, 760 Montauk Hwy., East Patchogue, 631-475-8288. Moderate. In a former Pizza Hut, savor the deliciously complex flavors of Thailand. Tom kha gai (chicken coconut soup) is not to be missed, and pad see yue, a chow fun-like noodle dish, excels. Staffers are helpful and hospitable, prices reasonable at both lunch and dinner.

The Bayport House, 291 Bayport Ave., Bayport, 631-472-2444. Expensive. The newest resident of The Bayport House is the best. Now, this is an exciting, new American restaurant with accents Asian, European and Latin. Go from hacked chicken with peanut sauce to a combo of blood sausage, wild mushrooms, polenta and manchego cheese; from veal Oscar to chicken in mole; angel hair in tomato-tinted seafood sauce to Thai-marinated hanger steak. All first rate.

Satelite Pizza, 799 Montauk Hwy., Bayport, 631-472-3800. Inexpensive. The stuffed pizzas are husky meals and a key reason to stop here. But the other pies, especially the Sicilian ones, are irresistible. And see if they have zeppoles, for your own private feast. Eat in or take out, you'll be happy.

BETHPAGE, PLAINVIEW

Ozumo, 164 Hicksville Rd., Bethpage, 516-731-8989. Moderate. Everything is vibrant at this cheerful Japanese restaurant whose menu features color photos of the dishes served. The sushi and sashimi are wonderfully fresh and flavorsome. Try the spicy tuna roll, the fish flecked with bits of fresh green chili. Appetizers like gyoza (dumplings) and tatsuta age (fried chicken) are executed with skill and elan. You'll also like the expertly fried tempura.

Pancho's Border Grill, 4119 Hempstead Tpke., Bethpage, 516-579-5500. Moderate. At this fine and funky Tex-Mex, walls are whimsically decorated with license plates from all over the United States; Grateful Dead music plays on the sound system. The perky cilantro-accented salsa is the first sign that this is not an ordinary chili joint. At lunch, try the catfish po' boy featuring cornmeal-crusted fish served with chipotle tartar sauce and terrific sweet potato fries. Fajitas are fine. So is the lush flan, served with freshly whipped cream.

Robata of Tokyo, 1163 Old Country Rd., Plainview, 516-433-5333. Moderate. At this Japanese restaurant, the sushi, which happens to be terrific, is upstaged by the meat, fish and vegetables prepared on the restaurant's unusual open-fire robata grill. Try the grilled shrimp, the wasabi chicken, the tsukune (chicken meatball) or kabocha (Japanese pumpkin) to find out why.

West Coast Kitchen, 377-2 S. Oyster Bay Rd. (in Plainview Centre), Plainview, 516-931-8300. Inexpensive. If a lively wrap and a yogurt-and-fresh-fruit smoothie are what you crave, head over to this friendly little order-at-the-counter eatery to find satisfaction. The Thai chicken wrap is a standout; so is the "unwrapped" sauteed chicken and shrimp in a creamy red pepper sauce, served with a good Caesar salad. In summer, eat at an umbrella table in the courtyard.

COLD SPRING HARBOR

Inn on the Harbor, 105 Harbor Rd., Cold Spring Harbor, 631-367-3166. Expensive. Attractive and appealing, especially in summer. The restaurant has a continental menu, with some fine French specialties, and a water view from the upstairs dining room. The cassoulet is excellent, as is duckling in orange sauce.

Wyland's, 55 Main St., Cold Spring Harbor, 631-692-5655. Inexpensive. A luncheon mainstay that has the look and feel of an eatery in New England, or the Midwest. The specials are worth noting. Good clam chowder, corn chowder, fish-and-chips, meat loaf and pies such as pecan and apple.

EAST HAMPTON

Della Femina, 99 N. Main St., East Hampton, 631-329-6666. Very expensive. The adman's namesake establishment is one of the East End's best, for the food and the parade of boldface customers. The fare is new American, changing regularly. Risotti, salads, seafood with Asian seasoning all star. Be prepared for crowds.

East Hampton Point, 295 Three Mile Harbor Rd., East Hampton, 631-329-2800. Very expensive. A lovely waterside locale is among the lures at this excellent new American restaurant. The seafood is especially recommended, whether raw or cooked. It's mobbed on summer weekends.

Rowdy Hall, 40 Main St., East Hampton, 631-324-8555. Moderate. A pub-bistro hybrid where you're tempted to become a regular. Casual style, good food. Recommended dishes include the fish and chips, made with cod; a first-class croque monsieur, the grilled ham-and-cheese sandwich; the Roquefort salad; onion soup; and the obligatory "Rowdyburger."

Santa Fe Junction, 8 Fresno Place, East Hampton, 631-324-8700. Moderate. Hidden away on a residential street is this terrific Southwestern restaurant, popular with locals and visitors alike. Ribs are slow-smoked and then baked to transcendental deliciousness, reason enough to come. Notable, too, are the cracker-crisp pizzas, the fine fajitas and the rich coffee-scented flan.

The Palm at the Huntting Inn, 94 Main St., East Hampton, 631-324-0411. Very expensive. The bucolic branch of the steak house chain. All the basics are first-rate, from the shrimp cocktail and crabmeat cocktail through the steaks and lobster, and sides such as creamed spinach and cottage fries. For dessert, cheesecake.

Turtle Crossing, 221 Pantigo Rd., East Hampton, 631-324-7166. Moderate. Hunker down for some serious barbecue at this cowboy-style Hamptons hangout. It's hard to imagine better eatin' than TC's jalapeño-studded hush puppies, deeply smoky pork ribs and chicken or definitive pulled-pork sandwich. Corn bread should not be missed.

ELMONT AREA

Cafe Zibibbo, 1007 Hempstead Tpke., Franklin Square, 516-326-9353. Moderate. The Italian fare is gently priced and the service affable at this pleasant trattoria. Start with the eggplant rollatine, but skip the puffy pizza whose toppings don't quite adhere. A good choice is the chicken and sausage topped with a colorful mix of roasted peppers. Conclude with black and white chocolate mousse cake.

Daniella, 127 Hempstead Tpke., Elmont, 516-326-2298. Moderate. The best bet in the neighborhood is this casual Italian eatery across the street from Belmont race track. The menu is limited, but much of it is very good. Stick with the day's specials, particularly pastas.

Fratelli Iavarone Cafe, 1534 Union Tpke. (Lake Success Center), New Hyde Park, 516-488-4500. Moderate. With a bustling pizza counter up-front (offering swell panini, too) and a busy full-service restaurant in the rear, this offshoot of the Iavarone Brothers Gourmet Market gets it right when it comes to pleasing a multigenerational clientele. Classics, like linguine with white clam sauce, are rendered with great care and skill, as are more adventurous specials. The only drawback is the wait for a table. It's a wait worth enduring.

FREEPORT

Hudson & McCoy, 340 Woodcleft Ave., Freeport, 516-868-3411. Expensive. The most dependable restaurant along the Nautical Mile. The seafood and the steaks are reliable choices. Lots of specials. The place can get pretty noisy, upstairs and downstairs.

Nautilus Cafe, 46 Woodcleft Ave., Freeport, 516-379-2566. Moderate. Among the newer "veterans" of the strip, Nautilus launches itself with steaks and seafood. In season, try stone crabs. Also, baked clams, broiled swordfish and a juicy porterhouse steak.

The Schooner, 436 Woodcleft Ave., Freeport, 516-378-7575. Moderate. The last stop on the strip, with waterside dining. It's a big, old-fashioned place, right for the entire family, with genial service and a sense of pace. Stick with simpler dishes: lobster cocktail, steamed lobster, plain finfish.

GARDEN CITY, UNIONDALE

Dominican Restaurant, 10-19 Front St., Uniondale, 516-292-5700. Inexpensive. Central American comfort food is what is served at this converted diner, where the welcome is warm and the decor decidedly pink. Must-haves include the sopa de res (beef soup), superior fried plantains and succulent pernil (roasted pork). The chef also makes a mean Cuban sandwich.

Victory Oyster Bar & Smokehouse, 860 Franklin Ave., Garden City, 516-739-7660. Moderate. Lots of good food at this buoyant eatery, whether you're here for the shellfish, the barbecue or the smoked fare. Pick at random and enjoy. Wash it all down with one of the many top beers.

Waterzooi, 850 Franklin Ave., Garden City, 516-877-2177. Moderate. A Belgian bistro, with an exceptional selection of brews, superior presentations of mussels, fries of a high order, and a variation on those waffles to remind you of the 1964 World's Fair. Distinctive and inviting.

Zucca, Roosevelt Field, Garden City, 516-739-7119. Moderate. This sleek mall-side cafe offers lively Italian fare. You have the option of ordering in either individual or family-sized portions, but know beforehand that the single-sized servings usually feed at least two. Veal Milanese was crisp and light, rigatoni with pesto especially herbal and rich.

GLEN COVE

Dolcetto, 48 Cedar Swamp Rd., Glen Cove, 516-609-9393. Expensive. There's a continental revival underway here, at an address that has hosted a U.N. worth of restaurants. The shellfish-filled crepe, tuna carpaccio, minestrone, mushroom risotto and whole baby chicken suit the place.

Gonzalo and Joe's American Cafe, 5 School St., Glen Cove, 516-656-0003. Moderate. This is old-fashioned American cooking at its most soothing. At a table covered in a red-checkered cloth, savor a soulful meat loaf, a savory soup, a gratifying pulled-pork sandwich with sweet potato fries. Macaroni and cheese? But of course. Finish with bread pudding and smile when the check comes.

La Pace, 51 Cedar Swamp Rd., Glen Cove, 516-671-2970. Moderate. Four-star Italian restaurant, with a grand dining room and a restless, creative kitchen. You'll find all the familiar dishes, as well as more unusual stuff. The pastas are superior: gnocchi agnolotti, ravioli. The menu is only a starter. The specials invariably are terrific.

Marra's, 1 School St., Glen Cove, 516-609-3335. Moderate. At this bright, informal, well-priced Glen Cove eatery, you'll find swell pastas, pizzas, salads and paninis, along with an assortment of entrees, some traditionally Italian, others straying into Asian and regional American territory. In fine weather, you can dine on the outdoor patio. That is, if you can find a table.

Riva Grill, 274 Glen St., Glen Cove, 516-674-9370. Moderate. Casual, coastal Italian specialties highlight this newcomer. Vivid seafood salad, swordfish carpaccio, bass with mascarpone cheese, whole grilled snapper, zuppa di pesce. For the carnivores: the sliced steak on arugula with tomatoes and potatoes.

Wild Fig, 167 Glen St., Glen Cove, 516-656-5645. Moderate. Pide, the Turkish version of a pizza, is a specialty at this bright Mediterranean restaurant whose brick oven also turns out marvelous breads. The lahmajun, flat and crisp, topped with spiced minced beef, is recommended. So is an entree called karniyarik, meat-stuffed eggplant "boats." House-made Turkish pastries and desserts satisfy.

GREAT NECK

Bocca di Rosa, 24 Middle Neck Rd., Great Neck, 516-487-9169. Moderate. In this gracious little trattoria, everything is prepared with skill and elan, from the freshly baked wheat bread to the house-made pasta (try the marvelous raviolini al burro e salvia). At lunch, salads are exceptional, as are panini on house-baked bread. Desserts are worth the caloric expenditure. The downside? Unbelievable noise levels on weekend evenings.

Matsuya, 6 Great Neck Rd., Great Neck, 516-773-4411. Moderate. A dependable Japanese restaurant, whether you're interested in sushi or cooked dishes. The broiled eel and the chicken-and-vegetable tempura are fine. So's the cool zaru soba, buckwheat noodles for dipping. The sushi counter excels with conical, special hand rolls.

Oyama, 148 Middle Neck Rd., Great Neck, 516-487-8460. Moderate. Plush upholstered armchairs make this Great Neck Japanese restaurant a lot more comfortable than most. Sushi is impeccable and presented with artistry. Try the black pepper tuna roll with avocado; it's spicy with a subtle hint of lemon. Cooked dishes, such as the chicken yaki soba, are a treat.

Peter Luger, 255 Northern Blvd., Great Neck, 516-487-8800. Very expensive. The suburban offspring of the Brooklyn classic. They share a singular devotion to red meat, and the porterhouse is, ounce-for-ounce, the best steak around. The bone-in sirloin also is recommended. Familiar side dishes, cheesecake to finish. But steak is all.

GREENPORT, CUTCHOGUE

Antares Cafe, 2530 Manhasset Ave., Greenport, 631-477-8839. Moderate. Ambitious kitchen, friendly staff and a modest water view define this new North Fork star. Some favorites: roasted saddle of lamb, roasted lobster with sweet wine sauce, butternut squash soup, Champagne-raspberry sorbet. Situated in the Brewer Yacht Yard.

Ile de Beauté, 314 Main St., Greenport, 631-477-2822. Moderate. Crepes are the highlights at this charmer. Start with a bowl of onion soup, and move on to one or more of the 100-plus crepes. The Roquefort production heads the savories. Enjoy brunch, too. The dining room has a vintage style. You'll want to linger.

The Greenport Tea Co., 119A Main St., Greenport, 631-477-8744. Moderate. The place does make you smile, courtesy of the cute decorations, the finger sandwiches and the high-tea scones. It's not exactly one of those Parisian classics, but the spot is fun.

The Wild Goose, 4805 Depot Lane, Cutchogue, 631-734-4145. Expensive. The menu changes regularly at this eclectic, very good restaurant. The accents are French and beyond. Winning choices include seared sea scallops with a coconut-cauliflower-artichoke puree and duck confit with red onion marmalade.

HAMPTON BAYS, YAPHANK AREA

Carmans River Inne, 450 Main St., Yaphank, 631-345-3302. Moderate. A popular spot with local businessfolk (especially those who work at the nearby government offices), this Yaphank favorite is half barroom, half dining area. Start with one of the old-fashioned homemade soups. The menu offers simple continental standbys, which are fine, if not overly creative. Portions are ample, service friendly.

Crazy Dog, 122 Montauk Hwy., Westhampton Beach, 631-288-1444. Moderate to expensive. Two restaurants in one: a casual eatery up front, and a more adventurous spot in the back. The latter is especially good, with new American cooking. Try the soft-shell crabs with blue Peruvian potatoes and corn relish, a deluxe lobster club sandwich, the smoked short ribs, and sugared doughnuts that are reminders of zeppoles past.

JT's Place, 26 Montauk Hwy., Hampton Bays, 631-723-2626. Moderate. A casual, year-round restaurant. The kitchen specializes in steaks and seafood, plus dishes such as barbecued pork chops, baby back ribs. Look for the daily specials, too.

HUNTINGTON

Brasserie 345, 345 Main St., Huntington, 631-673-8084. Moderate. Evoking Paris with its pressed-tin ceilings and bistro-style decor, this convivial Huntington brasserie (located in the rear of a municipal parking lot) offers traditional Parisian fare with a few American twists. The kitchen can do tuna tartare with mangoes as well as a fine, rich onion soup gratinée. At lunch, the confit of pork sandwich is a standout.

Kebabi Alem, 292 Main St., Huntington, 631-425-7654. Moderate. Orhan Yegen just may be the best Turkish chef in the metropolitan area. In his elegant, urbane, yet reasonably priced Huntington restaurant, the fare is sensational. Think you've tasted hummus before? Try Yegen's; it's silkier, more luxurious than almost any other. Mücver (zucchini pancakes) made with yogurt and feta, are a revelation of flavor and texture. Fish, grilled with great skill, is a marvel, as is the piquant adana kebab (hand-chopped, grilled spiced lamb). For dessert, order the voluptuous almond pudding sprinkled with pistachios.

Kozy Kettle, 366 New York Ave., Huntington, 631-547-5388. Inexpensive. A soup specialist, where you can eat in or take out. The choices change weekly. Among the consistent winners are the vegetable potage, split pea with ham, lobster chowder, corn chowder and chicken potpie.

Munday's, 259 Main St., Huntington, 631-421-3553. Inexpensive. A village landmark, this soda fountain-style eatery and Huntington institution offers well-executed fare-familiar luncheonette classics supplemented by a more upscale list of breakfast, lunch and dinner specials. Early in the day, try the lush French toast souffle; at lunch, the burgers are smoky, juicy and fine; for dinner, go for a blue-plate special, such as Cornish hen with barbecue sauce and sweet potato fries. House-baked pies are de rigeur for dessert.

Oaxaca, 385 New York Ave., Huntington, 631-547-1232. Inexpensive. Mexican food with full flavor and very modest prices. They make a lively ropa vieja, or shredded beef and onions in hot sauce, black bean soup, chiles rellenos, huevos rancheros and husky quesadillas.

Tortilla Grill, 335 New York Ave., Huntington, 631-423-4141. Inexpensive. Tex-Mex with personality marks this standout storefront. The chili keeps cooking and the tortillas come out in bulk. There's a lot of reliable food. But the chili is essential, doled out by a friendly crew.

LOCUST VALLEY

Barney's, 315 Buckram Rd., Locust Valley, 516-671-6300. Expensive. A cozy, autumnal dining room and a busy, year-round bar area give Barney's a split-level style. So the country inn can be noisy. The menu changes regularly. Winners over the years include the ample crab cakes, seared duck breast, and wild striped bass with a citric beurre blanc.

Brown's Tavern, 4 The Plaza, Locust Valley, 516-676-5793. Moderate. Small and cozy, this pubby eatery is under the aegis of Diane Kaiser, who hails from Scotland and once cooked at Kensington Palace. Her home-style specialties include such favorites as fish and chips for lunch and, at dinner (entrees come with soup, appetizer and dessert) a whole roast Cornish hen with an old-fashioned, sage-scented stuffing.

LONG BEACH AREA

Fresco Creperie & Café, 150 A E. Park Ave., Long Beach, 516-897-8097. Inexpensive. A crepe can be a satisfying and rewarding meal. At this spiffy little Long Beach eatery, under the same ownership as Steven's Pasta Specialties, the crepes are light and, whether savory or sweet, a real treat. Entree crepes come with salads, making for a complete and economical lunch or dinner. Solo diners can occupy themselves with any of the newspapers or magazines on the rack near the entryway.

Nick DiAngelo, 33 W. Sunrise Hwy., Merrick, 516-379-2222. Moderate. In a handsome dining room, you can dine on Italian fare, ordering either individual or family-sized portions. The chopped salad is a winner, as is the ultra-garlicky chicken scarpariello. Linguine with clam sauce is fine, too. Finish with the super nut-studded bread pudding with real whipped cream.

Paddy McGee's, 6 Waterview Rd., Island Park, 516-431-8700. Moderate. A New York-style seafood house, just east of the Long Beach bridge. There's a water view and an affection for fried oysters, steamed lobsters, crab cakes, pan-roasted scrod and a wedge of Key lime pie.

Station and Park, 3340 Park Ave., Wantagh, 516-781-3200. Moderate. Chef Jeffrey Baruch offers big, colorful plates of food in a big, colorful Art Deco setting. Whether at dinner or lunch, it's hard to go wrong with any of his eye-catching presentations. Standouts include the warm spinach salad with mushrooms, roasted garlic, pancetta and pine nuts and fresh basil linguine with grilled chicken, mushrooms, sun-dried tomatoes and arugula. The dark chocolate "bag" with berry mousse, whipped cream and fresh berries is the way to finish.

Ted Milan, 36 E. Park Ave., Long Beach, 516-670-0007. Expensive. This restaurant has deftly made the transition from steak house to new American grill. The tuna with wasabi sauce, cream of mushroom soup, pan-seared black bass and garlic ribeye steak are typical of the menu.

TKOPS 180, 180 W. Park Ave., Long Beach, 516-431-0044. Moderate. The casual offspring of The Kitchen Off Pine Street is a brisk and friendly spot for salads, wraps, sandwiches and some heartier stuff. The barbecued chicken and the sliced pork are very good.

MONTAUK

Bird on a Roof, South Elmwood Avenue, Montauk, 631-668-5833. Inexpensive. Much beloved by knowing locals and vacationers, this Montauk standby is both a beachwear boutique and a quaint breakfast and lunch restaurant. Whether you eat indoors or on the outdoor patio, you can enjoy delicious omelets, pancakes and sandwiches, served by a staff that cares.

Caswells, 17 S. Edison Ave., Montauk, 631-668-0303. Expensive. A countrified restaurant near the beach, with a very good new American menu. Rare yellowfin tuna with soba noodles, herb-crusted rack of lamb, and penne tossed with lobster and shrimp are typical, and commendable.

Dave's Grill, Montauk Harbor, Montauk, 631-668-9190. Expensive. The summertime crowd overflows at Dave's, where reservations are an adventure. This slice of a building has a kitchen that does very well with grilled seafood, fish-and-chips, clam chowder, and the "chocolate bag" dessert.

The Harvest on Fort Pond, 11 S. Emery St., Montauk, 631-668-5574. Expensive. The advice is simple: see the lighthouse and eat here. It's the main event in Montauk. Portions feed two or more. Lamb T-bones with polenta, whole snapper with sweet-sour sauce, lobster cakes with saffron sauce, and the seafood bruschetta are all fine. The reservations system is ridiculous.

NORTHPORT, GREENLAWN

252 Broadway, 252 Broadway, Greenlawn, 631-262-7200. Moderate. What looks like a simple roadhouse from the exterior turns out to be a cozy, romantic spot for rousing Italian cuisine, with a few side trips to Asia. Be sure to call ahead, for tables book up quickly. The restaurant is only open Wednesday to Sunday, something to keep in mind.

Intermezzo, 10-12 Fort Salonga Rd., Northport, 631-265-1212. Moderate. Whether you're craving a slice of "grandma" pizza or a full-scale lunch or dinner, this small, sleek-looking Italian restaurant has what it takes to satisfy. Stick to the printed menu, though, if you're looking to keep the meal economical, for specials tend to be costly — and prices aren't always listed on the blackboard.

Show Win, 325 Fort Salonga Rd., Northport, 631-261-6622. Moderate. Some of the most colorful, imaginative and delicious sushi rolls on Long Island can be found in this Japanese restaurant. The menu lists close to 50, and they're all winners. If you're in need of a restorative, try the ginger-infused clam soup.

OYSTER BAY

Canterbury Ales Oyster Bar & Grill, 46 Audrey Ave., Oyster Bay, 516-922-3614. Moderate. Theodore Roosevelt would love this clubby pub-restaurant, especially since much of the memorabilia decorating its walls relate to him. You can order from a daily list of oysters, and, most of the time, you'll find game on the list of specials, some of them a bit ambitious. But even a simple burger is done right. Sometimes, that's all you need.

Mill River Inn, 160 Mill River Rd., Oyster Bay, 516-922-7768. Expensive. A small, very pretty place, and a tough reservation to get. The American menu changes daily. It has taken a more conservative route in the past few years, with simpler preparations. But the food can be very good.

PORT JEFFERSON AREA

Costa de Espana, 9 Trader's Cove, Port Jefferson, 631-331-5363. Moderate. Spanish cuisine, with some big flavors and kitschy asides. Octopus Galician-style with paprika and olive oil, veal with lemon sauce, garlic soup and black bean soup, and paella with lobster are standouts.

Miller's Table, 90 North Country Rd., Miller Place, 631-331-4848. Moderate. In a timbered room with a potbellied stove and lots of fresh daisies, you can indulge in the simple, well-executed American cuisine of chef Erik Hart. Hot biscuits, flecked with herbs, arrive wrapped in a linen napkin inside a tin canister. Have the terrific Buffalo-style calamari and move on to such comforts as char-grilled short ribs with sun-ripened chili sauce. Be sure, at the beginning of the meal, to order warm cookies, served on a cookie sheet, melting and marvelous.

Salsa Salsa, 142 Main St., Port Jefferson, 631-473-9700. Inexpensive.

You'll sit on a stool by a metal counter and probably have to wait awhile for the chance to do so. But the rewards are fresh, delicious wraps and burritos, among the best the area has to offer. The machengo burrito — steak, eggs, and all kinds of spicy fixin's — is a winner. So is the chicken salad wrap, made with fresh greens and fruity salsa.

Tangerine, 4747-31 Nesconset Hwy., Port Jefferson Station, 631-331-7975. Moderate. Good Chinese fare at this spare but friendly establishment. Some of the better dishes are oysters in black bean sauce, Beijing duck, Sichuan dumplings in sesame sauce, and tangerine-spiked beef. The honeyed, sesame-covered banana is the right finale.

Thai Gourmet, 4747-25 Nesconset Hwy., Port Jefferson Station, 631-474-0663. Moderate. The vibrant flavors of Thailand are served by the hospitable staff of this tiny strip-mall eatery. Although you may have to wait for your table, you can inhale the fragrance emanating from the grill of the open kitchen. Recommended dishes? Just about everything on the menu.

PORT WASHINGTON

Diwan, 37 Shore Rd., Port Washington, 516-767-7878. Moderate. Among the more grandiose Indian restaurants in Nassau and Suffolk, Diwan also is a very reliable one. It has a full, traditional menu, plus a buffet at lunch. The samosas, or pastries with potatoes and peas, are fine starters. Lamb vindaloo is spicy and savory, and the tandoori productions of chicken and shrimp are smoky and moist. Tasty breads, too.

Pickles, 42 Main St., Port Washington, 516-767-8585. Inexpensive. Need a good pastrami sandwich? A nice, fat knish? A comforting bowl of chicken soup with matzo balls? Come to Pickles, a Port Washington deli that's both contemporary and soothingly old-fashioned. French fries are made a la minute, and they're marvelous. Roasted chicken? Juicy and delicious. Don't forget the sour pickles, which are on the house and puckeringly good.

Salvatore's Coal-Fired Brick Oven Pizza, 124 Shore Rd., Port Washington, 516-883-8457. Inexpensive. Superb coal-oven pizza is the main draw at Salvatore's in Port Washington, owned by relatives of the folks who opened Patsy's in Harlem back in 1929. Frank Sinatra was a regular there, and his photos and music fill the joint. Order a blistered, crisp-crusted pie oozing fresh mozzarella (made in-house) and topped with just enough tomato sauce. You'd be hard-pressed to do better.

RIVERHEAD

Lobster Roll / Northside, 3225 Sound Ave., Riverhead, 631-369-3039. Inexpensive. The offspring of the long-standing star of the Napeague stretch. Informal and fine for fried fish and, of course, the namesake roll, a toasted hot dog bun heaped with lobster salad.

Riverhead Grill, 85 E. Main St., Riverhead, 631-727-8495. Moderate. A time capsule diner, circa Ike's first term. It's homey and pretty good. Just stick with the staples: meat loaf, pot roast, corned beef, fresh ham. Avoid the seafood. Breakfast is dependable.

Spicy's, 225 W. Main St., Riverhead, 631-727-7281. Inexpensive. You're not here for the atmosphere. Spicy's is beyond genteel, closer to rundown. But the barbecue is zesty stuff, whether you're eating in or taking out. Ribs, chicken, chopped beef and pork, plus corn bread.

Sultan's Kitchen, 1077 Old Country Rd., Riverhead, 631-369-9766. Moderate. In an ambience that's part food market, part restaurant, you'll find the savory Turkish pastries called pides. The kebabs don't stint on the spices; chicken kebab is particularly flavorsome. At lunch, the place is often filled with local businessfolk as well as shoppers from the nearby Tanger outlets.

Tweed's Restaurant & Buffalo Bar, 17 Main St., Riverhead, 631-208-3151. Moderate to expensive. Owned by bison farmer Ed Tuccio (who raises his herd for breeding only), this popular Riverhead gathering spot, named for the Tammany Hall machine politician, is situated in the venerable J.J. Sullivan Hotel. In cozy Victorian surroundings, you can enjoy bison burger and bison steak, along with a wide variety of pub-style and New American items.

SAG HARBOR

Estia's Little Kitchen, 1615 Bridgehampton-Sag Harbor Tpke., Sag Harbor, 631-725-1045. Moderate to expensive. A small restaurant with a small menu often means a chef with control over the details that can make a meal. Colin Ambrose (who also owns Estia in Amagansett) has everything under control at this low-key, high-quality spot serving East Enders breakfast, lunch and dinner. Ambrose grows most of his own herbs and vegetables; seafood is locally caught, the meat and poultry top-quality.

Paradise Cafe, 126 Main St., Sag Harbor, 631-725-6080. Expensive. Creative and satisfying cookery, in a restaurant tucked behind the Paradise Books store. Pan-roasted guinea hen with rosemary, five-spice duck, prosciutto-wrapped scallops and lavender-scented crème brûlée are recommended.

Phao, 62 Main St., Sag Harbor, 631-725-0055. Expensive. Thai cooking that's more in tune with American tastes. The place can get noisy. Brisk, workmanlike service. Worth sampling: spring rolls, green mango salad, mussaman curry, steamed Chilean sea bass with ginger-plum sauce, shredded spicy chicken with red onion.

The Beacon, 8 W. Water St., Sag Harbor, 631-725-7088. Expensive. The water view is delightful — especially at sunset. The dining room also is a beacon for anyone looking for snappy monkfish with Savoy cabbage and fennel, steamed lobster, Sichuan-spiced duck confit and a lemony crepe-souffle.

ST. JAMES

Kitchen a bistro, 532 N. Country Rd., St. James, 631-862-0151. Moderate. Robert Dixon, chef-owner of this idiosyncratic gem of a restaurant, is a culinary artist. This explains why there is invariably a wait on weekdays and nights for a table and a necessity to reserve at least a month ahead on weekends. But it's worth deferring gratification, for Dixon's cuisine is not only innovative and delicious, it's amazingly reasonable in price. Whatever you order — whether it's feather-light gnocchi with fresh tomatoes, superb sake-marinated short ribs or robust pork medallions wrapped in bacon — you can only win.

Mirabelle, 404 N. Country Rd., St. James, 631-584-5999. Very expensive. Refined and first-class French fare, sometimes with an Asian aside. The ever-changing menu has its share of treats. The two-course production of seared duck breast and confit of duck leg is excellent. Warm oysters accented with ginger, and the almond-ginger tart are favorites.

Vintage, 433 N. Country Rd., St. James. 631-862-6440. Very expensive. A clubby steak house that's the very upscale offshoot of Manero's in Syosset. The steakery staples are here, but so are some diversions. Sweet-sour calamari, baked clams, plump crabcakes, the porterhouse for two, filet mignon, and steamed lobster are recommended.

SAYVILLE

Aegean Cafe. 35 Main St., Sayville, 631-589-5529. Inexpensive. At this friendly little Greek taverna they feed you well, and, if you're on a tight schedule, they get you out on time. Sandwiches — either veal or chicken souvlaki — are savory standouts. Dependability, not innovation, is the name of the game here.

Cafe Joelle, 25 Main St., Sayville, 631-581-4600. Moderate. One of Sayville's most popular spots, this small but sophisticated cafe offers a creative contemporary American menu in warm, unpretentious surroundings. If you're planning on dinner, it pays to call ahead; weekends can be especially hectic.

Cesare's, 45 Foster Ave., Sayville, 631-589-7775. Moderate. A soulful, southern Italian restaurant with spirit, consistency and a bit of kitsch. You're greeted by a life-size statue of a centurion. He'd be content with the lobster fra diavolo; pork chops with peppers, onions and vinegar; chicken scarpariello and pasta e fagioli.

SEA CLIFF, ROSLYN

Friend of a Farmer, 1382 Old Northern Blvd., Roslyn, 516-625-3808. Moderate. You'll feel as though you're in a Hudson Valley farmhouse at this rustic Roslyn retreat overlooking a scenic duck pond. House-baked breads and pies excel. Don't miss the superior chicken potpie and homey shepherd's pie.

Joanne's Gourmet Pizza & Pasta, 500 Glen Cove Ave., Sea Cliff, 516-671-7222. Moderate. In this cozy little contemporary cottage, you'll find more than just good pizza and pasta. Almost everything served is prepared with extra care. Soups, like the rich pasta e fagiole, are robust and hearty. Fresh vegetables are used liberally in almost every dish, and always with an eye to color and flavor. Desserts are house-made; order the bananas Foster with freshly whipped cream.

Tease, 1363 Old Northern Blvd., Roslyn, 516-625-4223. Expensive. New American with many themes stars at this adroit restaurant. Winning choices include pan-roasted halibut, roasted oysters with fennel and shallots, crabcakes with corn puree and aioli; and sweets such as apple-candied ginger cobbler and strawberry shortcake.

Tupelo Honey, 39 Roslyn Ave., Sea Cliff, 516-671-8300. Expensive. New American cooking with touches of Euro, Asian and Latin fare ripples through this eye-catching spot. The recommended dishes include sauteed foie gras with cherries, pineapple and frisée; rack of pork spiked with toasted cumin and lemon; and the roasted peach with a pit of Cabrales cheese.

SHELTER ISLAND

Olde Country Inn, 11 Stearns Point Rd., Shelter Island Heights,

631-749-1633. Moderate. An amiable bed-and-breakfast with a spirited kitchen. It has been here since the 1880s. You'll savor the lobster bisque, oysters mignonette, tuna tartare, Dover sole meunière, roast duckling with blueberry sauce and the crème brûlée.

Planet Bliss, 23 N. Ferry Rd., Shelter Island, 631-749-0053. Moderate. If you're lucky, you'll spot a deer or two from your table on the front porch of this rustic country retreat where the organic fare is hearty and delicious. Memorable meals here have included crabcakes with ginger tamari sauce, grilled lamb chops over red onion risotto, and vegetable lasagna. For dessert, try the warm raisin-studded bread pudding or the bittersweet chocolate mousse.

SMITHTOWN, KINGS PARK

Casa Rustica, 175 W. Main St., Smithtown, 631-265-9265. Expensive. One of the mainstays of Long Island's Italian restaurants. Excellent food with service to match. The whole fish baked in a salt crust is terrific. That goes for most of the seafood dishes. Also, hearty chicken scarpariello, cold antipasti, salmon carpaccio, pastas and risotti. Tiramisu and cheesecake are dependable finales.

H2O, 215 W. Main St., Smithtown, 631-361-6464. Expensive. Sporty and upbeat, this fish house is as lively for its socializing as for the seafood. Crabcakes, seared tuna, fried calamari, steamed lobster, and the '50s-style shore dinner complete the major food groups.

L'incontro, 25 Main St., Kings Park, 631-544-6426. Expensive. The scion of Casa Rustica in Smithtown, and definitely the best table in the neighborhood. Rigatoni alla Siciliana with tomatoes and eggplant, fennel-and-orange salad, curry-crusted tuna, tiramisu, and panna cotta all are commendable.

Shiki, 97 E. Main St., Smithtown, 631-366-3495. Moderate. A satisfying Japanese restaurant, informal and to the point. The showy hand rolls are very good, as is the fried soft-shell crab with tangy ponzu sauce. The playful "coconut sushi" is a fun dessert after a dinner of pristine raw fish.

SOUTHAMPTON

basilico, 10 Windmill Lane, Southampton, 631-283-7987. Expensive. The reigning Tuscan restaurant of the Hamptons, seasoned with celebrity appearances. Vitello tonnato, rigatoni with sausage and pink sauce, biscotti with sweet wine, loads of specials. Year after year, excellent.

George Martin, 56 Nugent St., Southampton, 631-204-8700. Expensive. The offshoot of the Rockville Centre eatery. An American-style "bistro" with good wine by the glass and very good dishes such as roasted chicken with cumin and candied garlic, dry-aged sirloin steak, marinated tuna, lobster roll, chocolate brownie sundae, and peach sorbet napoleon.

The Plaza Cafe, 61 Hill St., Southampton, 631-283-9323. Expensive. Exceptional seafood is the hallmark of the cafe, a good-looking spot that's very popular in season. The lobster-filled shepherd's pie, horseradish-crusted cod and sesame-crusted tuna are among the winners. The herbaceous rack of lamb will satisfy landlubbers.

STONY BROOK AREA

Pentimento, 95 Main St., Stony Brook, 631-689-7755. Moderate. Comfortable and consistent Italian eatery. It's a mix of old and new in the kitchen, deftly preparing grilled veal chops, sauteed duck breast, fried artichokes, fruit tarts and a host of tasty pastas.

Robinson's Tea Room, 99 E. Main St., Stony Brook, 631-751-1232. Inexpensive. Small, snug and veddy English, this cottage-like cafe, tucked into a Stony Brook shopping center, presents a welcome haven for the weary. Share the three-tiered tray of tea sandwiches, scones and pastries, along with a pot of freshly brewed tea. You'll also find a full range of salads, sandwiches, wraps and entrees. And no shortage of desserts.

Three Village Inn, 150 Main St., Stony Brook, 631-751-0555. Expensive. Traditional American cooking in a handsome setting defines the inn, which has been around since 1751. In the inn's current life, highlights include a full turkey dinner, generous prime rib, and a "tipsy Sherry trifle" for dessert.

Village Bistro, 766 Rte. 25A, East Setauket, 631-941-0430. Moderate. If you're looking for fresh and inventive fare at prices almost too low to be true, try this comfortable restaurant, a stone's throw from SUNY Stony Brook. Pork-stuffed spring rolls are particularly winning, as is the seared duckling breast with a dried cherry demi-glace. Do plan on dessert; everything's house-made and wonderful.

WESTBURY AREA

Benny's, 199 Post Ave., Westbury, 516-997-8111. Expensive. A four-star Italian restaurant, with seamless service and a superb kitchen. Outstanding pastas, seafood, chops. The veal chop showered with morels, salt-cod cakes in red sauce, caponata, and whole grilled pompano are exceptional. Perfect zabaglione to conclude.

Cafe Baci. 1636 Old Country Rd., Westbury, 516-832-8888. Moderate. Garlic permeates virtually every bite of the Italian cookery served up at this long-standing Westbury favorite whose decor offers a time trip back to the '80s. Pizza is always a good choice, and spaghetti with meatballs will sustain you for a long time.

PF Chang's China Bistro, The Mall at the Source, 1504 Old Country Rd., Westbury, 516-222-9200. Moderate. In the cavernous, theatrically decorated dining room of this corporate-owned Chinese restaurant, you'll find some surprisingly good food. The zippy hot and sour soup is a good way to begin. Spice-encrusted rare tuna over greens works well; so does dan dan (egg) noodles with a garlicky chicken sauce. A non-Asian but pleasing conclusion is the flourless chocolate cake.

The Palm Court, Carltun complex in Eisenhower Park, 516-542-0700. Very expensive. The dining room is handsome and the continental food generally good. But you'll effortlessly run up quite a tab. The Wine Cellar has clubby, and understandably vinophile, appeal.

The Stockyard, 103 Post Ave., Westbury, 516-876-6576. Expensive. A good steak house, friendly and informal. The prime rib is easily recommended, as is the veal chop. Order on the rare side, since they do tend to overcook things.

Thomas' Ham'n Eggery, 325 Old Country Rd., Westbury, 516-333-3060. Inexpensive. On any given morning — weekends especially — you'll find a line trailing out the door of this breakfast magnet (which also serves lunch and dinner). The reasons become clear as soon as you savor the ambrosial oatmeal, the fluffy pancakes, the terrific pecan waffles, the eggs and omelets served in skillets. House-made muffins and coffee cake excel. So does the welcome accorded regulars and newcomers alike.

Yuki's Palette, 611 Old Country Rd., Westbury, 516-334-5009. Moderate. The sushi here is as artfully presented as it is fresh. Try the spicy crunchy tuna roll or the unusual lobster roll. Tempura is ultra-crisp and greaseless. A big bonus is dessert: house-made chocolate mousse with freshly whipped cream. ◆

Long Island Wineries

Bedell Cellars, Main Road (Route 25), Cutchogue, 631-734-7537. **Web:** www.bedellcellars.com. **Hours:** 11 a.m. to 5 p.m. daily, except Christmas, Thanksgiving and Easter. **Special events:** Fall barrel-tasting weekends. Tours are limited to 10 or more and should be scheduled two days in advance.

Bidwell Vineyards, Route 48, Cutchogue, 631-734-5200. **Hours:** 11 a.m. to 6 p.m. daily through early September, then 11 a.m. to 5 p.m. **Special events:** Music on weekends, call for programs. Tours are by appointment.

Castello di Borghese-Hargrave Vineyard, Route 48, Cutchogue, 631-734-5111. **Web:** www.castellodiborghese.com. **Hours:** 11 a.m. to 5 p.m. Closed on Tuesdays January through March. **Special events:** Wine tastings, call for information. Tours are by appointment.

Channing Daughters, 1927 Scuttlehole Road, Bridgehampton, 631-537-7224. **Web:** www.channingdaughters.com. **Hours:** 11 a.m. to 5 p.m. Closed Tuesday to Thursday January through March. **Special events:** Call for details.

Corey Creek Vineyards, Main Road (Route 25), Southold, 631-765-4168. **Web:** www.coreycreek.com. **Hours:** 11 a.m. to 5 p.m. **Special events:** Concerts, call for schedule.

Duck Walk Vineyards, 132 Montauk Highway (Route 27), Water Mill, 631-726-7555. **Web:** www.duckwalk.com. **Hours:** 11 a.m. to 6 p.m. daily. **Special events:** Live music Saturday and some Sundays, call for schedule. Daily tours are conducted at noon, 2 p.m. and 4 p.m., with more added during summer.

Galluccio Estate Vineyards-Gristina Winery, Main Road (Route 25), Cutchogue, 631-734-7089 or e-mail info@gristinawines.com. **Web:** www.gristinawines.com. **Hours:** 11 a.m. to 5 p.m. Monday to Friday; 11 a.m. to 6 p.m. Saturday and Sunday. **Special events:** Call for schedule. Tours by appointment.

Jamesport Vineyards, Route 25, Jamesport, 631-722-5256 or e-mail JamesportVineyards@msn.com. **Web:** www.Jamesport-Vineyards.com. **Hours:** 10 a.m. to 6 p.m. daily or by appointment. During winter, open Tuesday and Wednesday by appointment. **Special events:** Jazz festivals, call for dates. Tours are informal and available on request.

Laurel Lake Vineyards, 3165 Main Road (Route 25), Laurel,

631-298-1420 or e-mail laurellake@llwines.com. **Web:** www.llwines.com. **Hours:** 11 a.m. to 6 p.m. **Special events:** Live music, call for details. Tours by appointment.

The Lenz Winery, Main Road (Route 25), Peconic, 631-734-6010 or e-mail pc@lenzwine.com. **Web:** www.lenzwine.com. **Hours:** 10 a.m. to 6 p.m. daily, June through October, 10 a.m. to 5 p.m. daily November through May. **Special events:** International tastings for chardonnay in July, merlot in September; call for dates, times and fees. Tours by appointment.

Lieb Family Cellars, 35 Cox Neck Rd., Mattituck, 631-734-1100 or e-mail Liebcellars@aol.com. **Web:** www.liebcellars.com. **Hours:** 11 a.m. to 6 p.m. daily (closed Mondays). During winter months, 11:30 a.m. to 5:30 p.m. Saturday and Sunday. Tours by appointment.

Macari Vineyards, 150 Bergen Ave., Mattituck, 631-298-0100 or e-mail macari@peconic.net. **Web:** www.macariwines.com. **Hours:** 11 a.m. to 5 p.m. daily; weekends only in winter. Tours by appointment.

Martha Clara Vineyards, 6025 Sound Ave., Riverhead, 631-298-0075 or e-mail info@marthaclaravineyards.com. **Web:** www.marthaclaravineyards. com. **Hours:** 11 a.m. to 6 p.m. daily. Free weekend tours by carriage.

Osprey's Dominion Winery, Main Road (Route 25), Peconic, 631-765-6188 or e-mail winemakr@ospreysdominion.com. **Web:** www. ospreysdominion.com. **Hours:** 11 a.m. to 6 p.m. daily; 11 a.m. to 5 p.m. in winter. **Special events:** Call or check the Web site for details. Tours are available on request.

Palmer Vineyards, 108 Sound Ave., Aquebogue, 631-722-9463 or e-mail palmervineyards@mail.com. **Web:** www.palmervineyards.com. **Hours:** 11 a.m. to 6 p.m. daily, June through October; 11 a.m. to 5 p.m. daily, November through May. **Special events:** Fall festivals featuring live bluegrass music and hayrides to vineyards, call for dates. Self-guided tours.

Paumanok Vineyards, 1074 Main Road (Route 25), Aquebogue, 631-722-8800 or e-mail mail@paumanok.com. **Web:** www.paumanok.com. **Hours:** 11 a.m. to 6 p.m. daily; in winter, 11 a.m. to 5 p.m. **Special events:** Jazz and classical concert series, harvest celebration and oysters and chenin blanc tastings; call for dates and times. Informal weekend tours on request during summer. Call in advance for weekday tours.

Peconic Bay Winery, Main Road (Route 25), Cutchogue, 631-734-7361 or e-mail info@peconicbaywinery.com. **Web:** www.peconicbaywinery.com. **Hours:** 11 a.m. to 5 p.m. weekdays, 11 a.m. to 6 p.m. weekends. Tours by appointment.

Pellegrini Vineyards, Main Road (Route 25), Cutchogue, 631-734-4111.

Web: www.pellegrinivineyards.com. **Hours**: 11 a.m. to 5 p.m. daily. Tours for individuals are self-guided. Guided tours for groups should be scheduled in advance.

Pindar Vineyards, Main Road (Route 25), Peconic, 631-734-6200. **Web**: www.pindar.net. **Hours**: 11 a.m. to 6 p.m. daily. **Special events**: Live music and tastings and Fall Harvest Festival Barbecue; call for schedules. Tours are at noon, 2 p.m. and 4 p.m.

Pugliese Vineyards, Main Road (Route 25), Cutchogue, 631-734-4057. **Web**: www.pug liesevineyards.com. **Hours**: 10:30 a.m. to 5 p.m. daily; 10:30 a.m. to 6 p.m. weekends.

Raphael, Main Road (Route 25), Peconic, 631-765-1100 or e-mail info@ raphaelwine.com. **Web**: www.raphaelwine.com. **Hours**: 12 p.m. to 5 p.m. daily.

Ternhaven Cellars, 322 Front St., Greenport, 631-477-8737. **Hours**: 11 a.m. to 6 p.m. Friday to Sunday (closed January through March); call for extended summer hours. Tours are available.

Wolffer Estate, 139 Sagg Rd., Sagaponack, 631-537-5106 or e-mail info@ wolffer **Web**: www.wolfer.com. **Hours**: 11 a.m. to 6 p.m. daily, May through October; 11 a.m. to 5 p.m. daily, November through April. Group tours by appointment. ◆

Newsday Photo / Bill Davis

A busy day at the tasting bar at Pindar Vineyards in Peconic

More Destinations

F OLLOWING is a list of Long Island sites not previously mentioned in this book that day-trippers also might find interesting.

STATE PARKS

Shadmoor State Park, Route 27 and Seaside Avenue, Montauk; 631-668-5000. This 99-acre oceanfront park (owned jointly by New York State, Suffolk County and East Hampton Town, which co-manages it with The Nature Conservancy) opened in 1991. It has hiking and biking trails plus two bunkers built during World War II for surveillance (and thus designed to look like cottages from the water). Open daily year-round; free.

State Park at Camp Hero, next to Montauk Point State Park, Montauk; 631-668-5000. This 415-acre former military base was expected to be officially named and opened to the general public as New York's newest state park in May or June of 2002, with hiking, biking, bird-watching, surf casting and displays pertaining to the site's Cold War history as a radar defense station (portions will be off-limits during continuing cleanup work). Fees and hours were to be determined.

NASSAU COUNTY PARKS

Resident passes are required for Nassau parks; call 516-572-0200.

Cedar Creek Park, Seaford (Wantagh Parkway Exit W6, east on Merrick Road), 516-571-7470; 259 acres with a bicycle path to Jones Beach, jogging course, roller-skating area, picnic areas, ball fields and courts (some requiring a fee and / or permit).

Centennial Park, Roosevelt (Southern State Parkway Exit 21, south on Nassau Road to Centennial Avenue), 516-571-8695; two acres with playground, ball courts, table games.

Inwood Park, Inwood (Southern State Parkway Exit 19, Peninsula Boulevard south to Nassau Expressway 878 south, Bayview Avenue west), 516-571-7894; 16 acres with ball fields and courts, playground, outdoor roller rink, saltwater fishing, launch ramp (annual fee), picnic area, walking trail.

North Woodmere Park, North Woodmere (Southern State Parkway Exit 19, Peninsula Boulevard south to Branch Boulevard north to Hungry Harbor Road west), 516-571-7801; 150 acres with lighted ball fields and courts (some fees), five pools (fee), nine-hole golf and driving range (fee), fishing.

The Rev. Arthur Mackey Sr. Park (formerly Roosevelt Park), Roosevelt (Southern State Parkway Exit 23, south on Meadowbrook Road to Washington Avenue, west to Lakeside Drive), 516-571-8692; 27 acres with nature trail, bicycle trail, ball field and courts, fishing, picnic areas.

Wantagh Park, Wantagh (Wantagh Parkway Exit W6, west on Merrick Road), 516-571-7460; 111 acres with five pools, bicycle paths, picnic areas, fishing pier, boat ramp, marina, ball fields and courts (some requiring a fee and / or permit).

Whitney Pond Park, Manhasset (Long Island Expressway Exit 33, Community Drive north to park on the right, just before Northern Boulevard), 516-571-8300; 24 acres with ball courts, two pools and diving pool (fee), picnic areas, jogging course, fishing.

SUFFOLK COUNTY PARKS AND PRESERVES

Cedar Point County Park, East Hampton (Montauk Highway east, north on Stephen Hands Path, Old North West Road and Alewive Brook Road), 631-852-7620; 608 acres with self-guided nature trail, saltwater fishing (permit), rowboat rentals, picnic area, campsites.

Goldsmith's Inlet County Park, off Mill Road, Peconic, 631-854-4949. The 60-acre park is popular with bird-watchers.

Hoyt Farm Park Preserve, New Highway, Commack, 631-543-7804. This wooded, 133-acre preserve includes the Hoyt-Wicks Homestead Museum. Maple sugaring sessions are conducted in February and March.

Inlet Pond County Preserve, Main Road, Greenport, 631-854-4949. Features 45 acres on Long Island Sound. Popular for surf fishing.

Lake Ronkonkoma County Park, Lake Ronkonkoma (Long Island Expressway Exit 58, Nichols Road north, Smithtown Boulevard east; seasonal parking fee), 631-854-9699; 223 acres with ball fields, fishing.

Northwest Harbor County Park, off Swamp Road, East Hampton, 631-854-4949. This 337-acre preserve offers views of migratory waterfowl.

Orient Point County Park, end of Route 25, Orient, 631-854-4949. This 48-acre preserve is popular with beachcombers.

Poxabogue County Park, Old Farm Road, north of Montauk Highway and east of Bridgehampton, 631-854-4949; 26 acres, nature walk.

Smithtown Greenbelt County Park, off Route 347 at Nissequogue River, Smithtown, 631-854-4949. This 118-acre preserve is part of the Greenbelt Trail.

Suffolk Hills County Park, off County Road 51, Southampton, 631-854-4949; 985-acre preserve with Bald Hill, East End's highest point.

Photo by Gordon M. Grant

Hikers follow the shore at Northwest Harbor County Park, East Hampton.

Van Bourgondien County Park, West Babylon (Sunrise Highway Exit 38, south on Little East Neck Road, west on Arnold Avenue to Albin Avenue), managed by Babylon Town, 631-893-2100; 19-acre park with soccer fields, tennis courts, playground, exercise course.

MUSEUMS

Corwith Windmill, Village Green, Montauk Highway, Water Mill, 631-726-5984. Built in Sag Harbor in 1800 and relocated to Water Mill in 1813. Tours are available year-round by appointment.

1901 Restored Depot, South Broadway and South Third Street, Lindenhurst, 631-226-1254. This former depot was Suffolk County's first railroad museum. Call for summer concert dates.

Old Village Hall Museum, 215 S. Wellwood Ave., Lindenhurst, 631-957-4385. Village history is depicted in exhibits on home, industry and recreation.

Seaford Historical Society and Museum, Waverly Avenue, Seaford, 516-826-1150. Originally a schoolhouse and later a fire department headquarters, this 1893 site houses various historic displays.

Suffolk County Police Museum, 30 Yaphank Ave., Yaphank, 631-852-6174. Self-guided tours 10 a.m. to 4 p.m. daily. The museum traces the evolution of policing, from sheriff in the early 1600s to present-day police officer.

Telephone Pioneer Museum, 445 Commack Rd., Commack, 631-543-1371. Open to public only first Sunday of the month from 1 to 4 p.m. Among the displays is a re-creation of Alexander Graham Bell's workshop. ◆

Index

INDEX

INDEX

About the Author

Staff writer Barbara Shea has covered a variety of subjects for Newsday, most recently travel, and also has taught editing and feature writing at New York University. She has a master's degree from Syracuse University and a bachelor's from the University of Rhode Island. While she was an undergraduate, she began her journalism career as a reporter for The Providence Journal.

Acknowledgments

Every Newsday staffer listed at the bottom of the page made an indispensable contribution to this book, which was the idea of managing editor Howard Schneider. I am especially grateful to him for the assignment. Extra thanks also are due assistant managing editors Phyllis Singer, for setting the high standard, and Robert Eisner, for establishing the overall design; project editor Bob Henn, for his incomparable wit and wisdom day in, day out; news editor Lawrence Striegel and the copy editing staff, for the outstanding chapter headings and efforts to ensure accuracy; photo editor Tony Jerome and the other photographers whose work showcases the sites; food writers Peter Gianotti and Joan Reminick, for their incisive restaurant reviews and for their own informative books. It is my very good fortune to have worked with them all.

The following friends and colleagues who weren't directly involved in the production of the book also offered much appreciated tips, insights, encouragement and/or companionship on various excursions: Francine Brown, Sylvia Carter, Paula Hartman Cohen, Phyllis Gates, James T. Madore, Iris Quigley, Marjorie Robins, Ginger Rothe, Irene Virag and Virginia Wilson.

Special recognition is reserved for the dedicated staffers and volunteers of Friends for Long Island's Heritage; the Society for the Preservation of Long Island Antiquities; the Long Island Greenbelt Trail Conference; The Nature Conservancy; the National Park Service; the U.S. Fish and Wildlife Service; the parks departments of New York State, Nassau and Suffolk counties, as well as the innumerable historical societies and other public and private agencies and organizations that manage Long Island's many tourist attractions. Their knowledge and help made writing this book easier and more fun for me; their anecdotes and enthusiasm make visits so enjoyable for everyone. — **Barbara Shea**

Staff

Bob Henn, editor; Phyllis Singer, executive editor; Lawrence Striegel, news editor; Tony Jerome, photo editor; Richard Cornett, cartography; Jack Millrod, supervisory editor; Robert Eisner, book design; Richard Loretoni, cover design; Bob Newman, logo design; Peter Gianotti and Joan Reminick, restaurant reviews; Gary Dymski, Ronnie Gill, James Stephen Smith, Richard L. Wiltamuth, copy editing and layout; Diane Daniels, Virginia Dunleavy, Elayne Feld, Leigh Anne Fields, Daniela Hanson, Alice Norkett, Lynn Petry, Pat Sollitto, Gene Sullivan, research; Dorothy Levin, Laura Mann, Eileen Effrat, indexing. Imaging by Newsday Prepress Department.

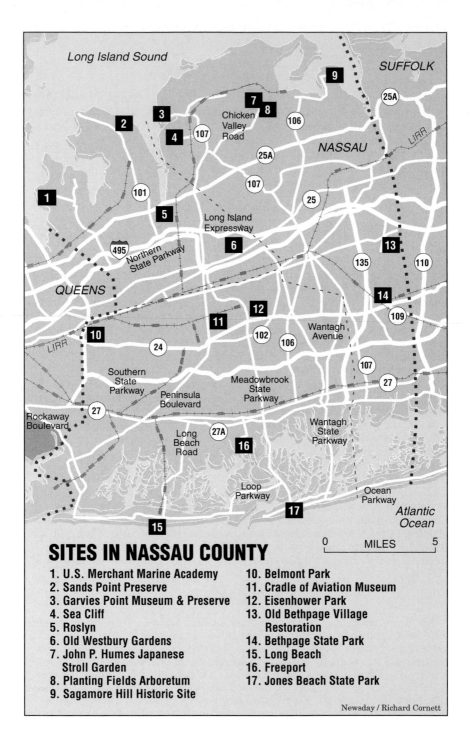

SITES IN NASSAU COUNTY

1. U.S. Merchant Marine Academy
2. Sands Point Preserve
3. Garvies Point Museum & Preserve
4. Sea Cliff
5. Roslyn
6. Old Westbury Gardens
7. John P. Humes Japanese Stroll Garden
8. Planting Fields Arboretum
9. Sagamore Hill Historic Site
10. Belmont Park
11. Cradle of Aviation Museum
12. Eisenhower Park
13. Old Bethpage Village Restoration
14. Bethpage State Park
15. Long Beach
16. Freeport
17. Jones Beach State Park

Newsday / Richard Cornett

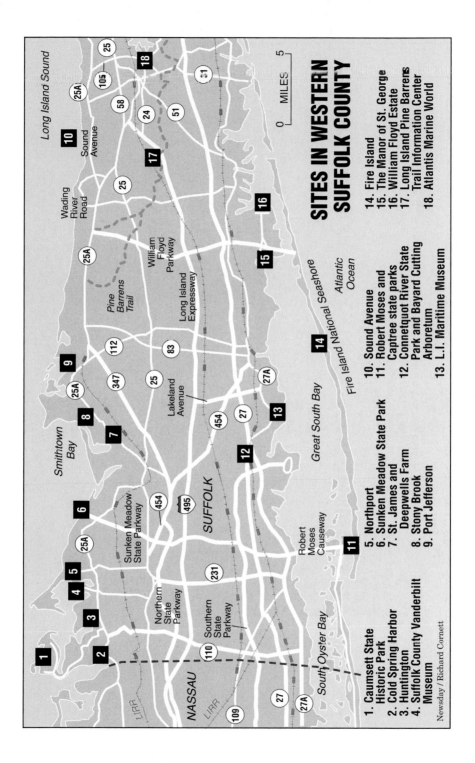

SITES IN WESTERN SUFFOLK COUNTY

1. Caumsett State Historic Park
2. Cold Spring Harbor
3. Huntington
4. Suffolk County Vanderbilt Museum
5. Northport
6. Sunken Meadow State Park
7. St. James and Deepwells Farm
8. Stony Brook
9. Port Jefferson
10. Sound Avenue
11. Robert Moses and Captree state parks
12. Connetquot River State Park and Bayard Cutting Arboretum
13. L.I. Maritime Museum
14. Fire Island
15. The Manor of St. George
16. William Floyd Estate
17. Long Island Pine Barrens Trail Information Center
18. Atlantis Marine World

Newsday / Richard Cornett

319

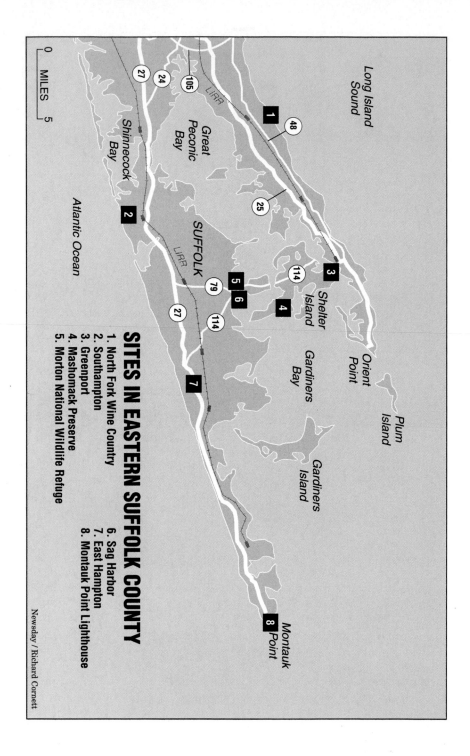

SITES IN EASTERN SUFFOLK COUNTY

1. North Fork Wine Country
2. Southampton
3. Greenport
4. Mashomack Preserve
5. Morton National Wildlife Refuge
6. Sag Harbor
7. East Hampton
8. Montauk Point Lighthouse

Newsday / Richard Cornett

320